PORSCHE 911

ICONS OF EXCELLENCE

PEFC/10-31-1895

Editions Glénat
Couvent Sainte-Cécile
37 rue Servan
38000 Grenoble
www.glenat.com
contact@glenat.com

French edition:
Text: Sylvain Reisser
Editing: Coline Rouge
Graphic adaptation and layout: S.B.

Prepress and manufacturing: Glénat Production

Legal deposit: May 2025
FRENCH ISBN: 978-2-344-06650-8 / 001

Printed in France in 2025 by Imprimerie Chirat,
on paper from sustainably managed forests

English edition:
Editorial direction and translation: Glyn and Jean Morris
Dalton Watson Fine Books
Deerfield, IL 60015 USA
www.daltonwatson.com

Graphic adaptation: Jodi Ellis
Prepress and manufacturing: Glénat Production

USA ISBN: 978-1-956309-31-7

Cover photograph:
The 1998 911 GT1, one of the most spectacular models in history with a 3.2-liter 544 hp flat-six turbo. Only 36 units were produced for road and race. The car was also homologated for endurance in the GT1 category.

PORSCHE

911

ICONS OF EXCELLENCE

SYLVAIN REISSER

FOREWORD BY
JÜRGEN BARTH

DALTON WATSON FINE BOOKS

FOREWORD

This book, which brings together all of the most exceptional Porsche 911 models, is a captivating work. For the first time, it offers a unique perspective on the legendary rear-engined sports car, tracing its sixty-two-year history through every special – and most often limited – edition.

I began my apprenticeship at Porsche in 1963 as an automotive locksmith and subsequently pursued additional training as a salesperson. After these first two positions, I worked under the direction of Huschke von Hanstein in the press department. Consequently, I had the opportunity to contribute to the development of numerous models and to drive them in races or rallies.

I still vividly recall my first encounter with the 911 R in 1969 when I was responsible for Porsche's race management and rally commitments. I had to oversee Gérard Larrousse's participation in the Tour de France, which he won at the wheel of a 911 R. Huschke von Hanstein wanted to present the trophy to Ferry Porsche on Monday morning, so I drove all night, from Bordeaux to Stuttgart. Fortunately, there were no speed restrictions in France at the time, which allowed me to drive at full speed.

Later, I had the chance to drive the 911 Carrera 2.7 RS, followed by the 3.0 RSR, with which I won several 1000 km races. At that time, I was also responsible for homologating the cars, so I knew these vehicles inside and out.

Finally, in 1978, while I was team manager for Porsche-Martini, I participated in the Safari Rally. My in-depth knowledge of 911 technology helped me guide our cars to the finish line.

I hope you enjoy this captivating book.

Jürgen Barth
Former Porsche driver, winner of the 1977 24 Hours of Le Mans
with Jacky Ickx and Hurley Haywood

PREFACE

Since its presentation at the 1963 Frankfurt Motor Show, and after seven years of development, under the designation 901, the Porsche 911 has been part of the landscape. However, it has not always had an easy life, narrowly avoiding being superseded on several occasions. The appearance of the 911 Turbo in 1974 helped reinforce its appeal.

After addressing its early shortcomings, the 911 has continued to evolve while consistently adhering to its original design principles. Born with a 2-liter flat-six engine delivering a modest 130 hp, this 2+2 coupe, more bourgeois than sporty, managed, through continuous improvement, to evolve the rear-engine architecture that seemed, in many respects, outdated.

Today, no one could imagine it not being part of the landscape. The 911 has become one of the most desirable cars ever made, and its prestige has only continued to increase despite fierce competition. Without competition, the 911 certainly would not have had the same destiny. This is what the book aims to show.

There are several versions of 911s. The aim here is to highlight those that have an extra feature, either because their production was limited, or they were designed for racing, or they have sporty characteristics.

The sporty variants, developed in parallel with series production cars, also served as a basis for customer competition models. Some have distinguished themselves in events as varied as the 24 Hours of Le Mans and major international rallies on African tracks and on snow, attesting to the success of the initial concept. Conversely, racing experiences have allowed road models to be perfected. The reader will find evidence of this here, and the adventure, which began in 1963, is far from over.

Ferry Porsche, the founder of the brand bearing his name, often took mischievous pleasure in proclaiming that the last car built would be a sports car. What if it were the Porsche 911?

Sylvain Reisser

S-H 7945

I

TYPE F

1963–1973

Tasked with replacing the 356, which gave birth to the Porsche brand in 1948, the 911, introduced in 1963, established the premise of a Grand Touring car that was more spacious, more comfortable, and more powerful. After initial difficulties with road holding issues, these qualities became evident in competition across a multitude of terrains. History was in the making. The production of the first generation, which spanned ten years, reached 111,995 units.

"This car has no future." It was with this shocking phrase that Helmuth Bott, Porsche's prototype testing manager, concluded his report after testing the prototype of the 911 on November 1, 1960. The German manufacturer had launched the project to replace the 356, which was reaching the end of its capabilities, in the mid-1950s. The specifications for the new car stemmed from evolving driver expectations and the desire to attract a new clientele. According to the wishes of Ferdinand Anton Ernst Porsche, known as Ferry Porsche, the future vehicle had to be more spacious and offer a larger trunk, which required a longer wheelbase (but not exceeding 2.20 m), more comfortable, and more powerful. The sales department then insisted on the need to preserve the lines of the 356. This constraint posited an evolution rather than a revolution and implied retaining the architecture of a rear-mounted flat engine.

After several preliminary studies, including an investigation of a four-seater, the successor to the 356 began to take shape in the second half of the 1950s. External engineers and designers submitted several proposals, but none received approval from Ferry Porsche. Ultimately, the design by his eldest son, Ferdinand Alexander, known as "Butzi," came closest to the specifications. Presented in 1959, the study known by the code name "754 x T7" was reworked several times to result in the "T8" coupe project with 2+2 architecture. As for engineer Bott's diatribe, it did not go unnoticed within the Zuffenhausen design offices. It was less about the car itself than about the engine, the famous four-cylinder double overhead camshaft engine by engineer Ernst Fuhrmann. Under the management of Ferdinand Karl Piëch (Ferry Porsche's nephew) and Hans Mezger, the in-house engine designer, the management launched the design of an air-cooled flat-six engine, with an axial fan and a single overhead camshaft (SOHC) for each bank of three cylinders. With a displacement of 2 liters, the first version of this flat-six engine, featuring dry sump lubrication and fed by six single-barrel Solex carburetors, delivered 130 hp at 6,100 rpm and 175 Nm (129 lb-ft) of torque at 4,200 rpm. The 911 introduced a new five-speed synchronized manual gearbox and a limited-slip differential produced by ZF Friedrichshafen.

Bearing the type 901, which corresponded to the study number in Porsche's nomenclature, the prototype left the Zuffenhausen factory gates on November 9, 1962. Less than a year later, on September 12, 1963, the new 901 was the attraction of the Frankfurt Motor Show. While its basic design was final, the car was not yet ready for production. An additional year was needed to perfect its development. As was customary at Porsche, some of the thirteen development prototypes were given charming nicknames by the mechanics: petrel, bat, blue tit, lemon, bright blue, and Barbarossa. Following the German motor show, another issue attracted attention. Peugeot had industrial ownership of all three-digit numbers with a zero in the middle, based on a claim made in 1929. Porsche had no choice but to rename the 901, in the middle of its launch campaign. After careful consideration, the German firm decided, on November 22, 1964, to replace the 0 with a 1. A pragmatic choice: all commercial and technical documents were almost ready, as were the logos for the rear and the glove box lid.

RIGHT: As on the 356, the ignition key is to the left of the steering wheel but the finish is more refined. Over the years and evolutions, the layout of the instrumentation will remain unchanged.

PREVIOUS TWO PAGES: At the beginning of its history, the 911 changed with every new model year.

The Type 911 was born. The large luggage compartment under the front hood led to the adoption of transverse wishbones and MacPherson struts. At the rear, longitudinal swing arms were chosen for better guidance. With a wheelbase of 2,211 mm (87 in), the 911 was a very compact car, barely 4,163 mm (163.9 in) long and 1,610 mm (63.4 in) wide. It was homologated at 1,080 kg (2,381 lbs).

Although it ensured continuity of style and technology with the 356, the 911 was met with a cool reception. Complaints were numerous: a high price, mediocre power and performance, imperfect carburetion, and above all, approximate road holding. In addition to truly narrow tires – 165 HR 15 tires mounted on wheels only 4.5 inches wide – the 911 suffered from the inherent flaws of a rear-engine overhang architecture: oversteer, unstable directional stability, and high sensitivity to crosswinds.

When engineers failed to find a lasting solution, they reinforced the front bumper with two bumperettes, each weighing 11 kg (24 lbs).

Porsche also had to face a revolt from customers disappointed by the disappearance of the convertible. For them, barely a year after the commercial launch of the 911, a new Targa version, with a stainless steel roll bar allowing open-air driving with a very high level of safety, was introduced.

In parallel with the appearance of the first competition versions, engineers focused on evolving the range.

For the 1966 model year, triple-barrel Weber carburetors replaced the Solex equipment. The following year, the range was expanded with a 160 hp S version. Top speed increased from 210 (130 mph) to 225 km/h (140 mph). Five-inch light alloy wheels appeared; front and rear tracks were widened. For the first time, Porsche offered a Sportomatic four-speed semi-automatic gearbox as an option.

With each model year, the 911 underwent its share of evolutions. Thus, successively, Porsche's sports car received a downgraded 110 hp T version, mechanical injection, a wheelbase extended by 58 mm (2.28 in) to 2,268 mm (89.29 in), new chassis settings, and ventilated disc brakes. Displacement was increased to 2.2 liters with the 1970 model year: power outputs were 125, 155, and 180 hp. Two years later, the flat-six reached 2.4 liters. Power outputs increased slightly: 130, 165, and 190 hp. Soon, the question of the 911's future would arise again, but for entirely different reasons than those stated by engineer Bott.

911 R
1967

Following the delivery of the first 911s to its customers, the Zuffenhausen factory prepared a car for participation in the Monte Carlo Rally in January 1965. Entrusted to Peter Falk and Herbert Linge, the almost production-ready car finished fifth. This performance encouraged the competition department to go further. By applying the recipes reserved for competition models – a vitamin and weight-loss cure – the 911 should be able to nurture higher ambitions. Thus, at the 1967 Frankfurt Motor Show, Porsche launched a competition version developed from the 2-liter 911 and named "R" for Racing. The weight was reduced to 830 kg (1,830 lbs) by removing numerous comfort features and using aluminum for door and hood hinges, as well as plastic for the doors, hoods, front fenders, and both bumpers. With the exception of the 4 mm (.16 in) thick glass windshield, all windows were replaced with Plexiglas. The rear quarter windows received vents. Chassis modifications were minimal, but slightly wider fenders distinguished the R to accommodate six-inch wheels at the front and seven-inch wheels at the rear.

Symbol of the capabilities of the 2-liter engine borrowed from the 906 prototype, the 911 R's tachometer is graduated to 10,000 rpm.

The 2-liter 901/22 boxer engine came from the 906. The crankshaft ran on eight bearings, the cylinder heads were aluminum, and the cylinder liners were chrome plated. A dry sump system provided lubrication, and the two triple-barrel Weber 46 J carburetors were fed by two electric pumps. Ignition relied on two spark plugs per cylinder. Result: power peaked at 210 hp at 8,000 rpm.

Illustration of technology transfer between competition and road: the 2-liter flat-six comes from the Carrera 6 sports prototype. This engine with dual ignition and titanium connecting rods produces 210 hp.

Its silhouette was as discreet as its track record. Due to homologation in the "prototype" category, the 911 R did not have the expected career. Only 20 units were produced by the Stuttgart coachbuilder, Karl Baur, to which four prototypes built under Ferdinand Piëch's supervision must be added.

ENGINE: Flat-six
DISPLACEMENT: 2.0L / 121.5 in³
POWER: 210 hp at 8,000 rpm
TORQUE: 152 lb-ft at 6,000 rpm
GEARBOX: Five-speed
DIMENSIONS (L x W x H): 13' 7.9" x 5' 3.4" x 4' 2.4"
WEIGHT: 1,830 lbs
TOP SPEED: 149 mph

Homologated only in the Prototype category due to the low number of units produced, the 911 R is distinguished by its feather-light weight and its slightly widened fenders to accommodate larger wheels.

911 2.2 ST/2.3 ST
1970

Porsche benefited from a stroke of luck when the 2.2-liter engine 911 was introduced for the 1970 model year (starting August 1969), coinciding with a change in sporting regulations. Excluded from Group 2 (Touring), the Zuffenhausen sports car was now homologated in Group 3 (Grand Touring). Engineers largely drew inspiration from the 911 R to develop the 2.2 ST and 2.3 ST versions. The flat-six approached that of the Carrera 6 with dual ignition, competition camshafts, and pistons. A Bosch mechanical injection system handled fueling with a three-dimensional cam. The engines of 911s homologated in Group 4 could be equipped with two Weber 46 IDA3 triple-barrel carburetors, without affecting power.

Compared to the 2.2 S, with which it shared the same displacement, the 2.2 ST gained 50 hp (230 hp) because of specific tuning and some modifications. Permitted by regulations, the displacement was allowed to increase to 2,380 cm³ a few months later, and the 911 2.3 ST produced 250 hp. The increase in performance was also based on a series of measures: reduction of mass due to the removal of all interior comfort elements, adoption of aluminum and plastic body elements, lowered chassis and stiffer shock absorbers, wider wheels and tires (seven inches at the front and nine inches at the rear).

Unlike the 2.2 ST, which was primarily intended for road events, the 2.3 ST quickly became the preferred GT for customers participating in circuit races. Regardless of the version, these two competition 911s won numerous victories during the 1970 and 1971 seasons. The most renowned remains the 2.2 ST's one-two finish at the 1970 Monte Carlo Rally.

ABOVE: In 1970, Porsche achieved its third consecutive victory, the second for the Björn Waldegård-Lars Helmer crew, in the Monte Carlo Rally with the 911 S 2.2 ST entered by the team.

RIGHT: Under a silhouette almost identical to the production model, the 2.2 ST used lightweight materials, while the flat-six employed solutions previously reserved for its endurance prototypes.

LEFT: The interior treatment of the 2.2 ST and 2.3 ST competition versions differed little from that of the 911 R. The objective was weight reduction.

ENGINE: Flat-six
DISPLACEMENT: 2.2L / 2.4L (137.1 in / 145.2 in)
POWER: 230 hp / 250 hp at 7,800 rpm
TORQUE: 170 lb-ft / 187 lb-ft at 6,200 rpm
GEARBOX: Five-speed
DIMENSIONS (L x W x H): 13' 7.3" x 5' 3.4" x 4' 2.4"
WEIGHT: 1,852 lbs
TOP SPEED: 149 mph

911 2.5 ST
1972

As with the 2.2 and 2.3 ST, the engineers focused on lightening the interior by using lightweight felt and bucket seats.

In the early 1970s, competition 911s were constantly evolving. In preparation for the 1970 Tour de France, Porsche prepared an exceptional machine, which remains a unique piece to this day, as only one example was produced. Entrusted to French driver Gérard Larrousse, the 911 competed in the "Prototype" category, for good reason: it had little in common with other competition 911s. The Essonne-based tuner Louis Meznarie, who was entrusted with the mechanical preparation, managed to extract 256 hp from the flat-six with triple-barrel Weber carburetors, pushed to 2,395 cm^3. The power-to-weight ratio set a new record. Engineers reduced the weight to 780 kg (1,720 lbs) by meticulously shaving off gram by gram wherever possible. Since then, no 911 has done as well. The crew of Gérard Larrousse and Maurice Gélin drove the psychedelically decorated yellow and orange 911 to third place in the major event, behind the two Matra prototypes.

In parallel, Porsche's competition department took advantage of the introduction of the 2.4-liter 911 S to develop a 2.5 ST version. As with all models destined for racing, the team in charge of studying the new vehicle focused its research on power, weight, and road holding. The displacement, increased to 2,492 cm^3, allowed for a power output of 271 hp at 7,900 rpm and a torque of 265 Nm (195 lb-ft) at 6,200 rpm. This long-stroke engine proved to be less reliable than its predecessor due to crankshaft vibrations. According to factory records, only seven examples of the 911 2.5 ST were produced.

ENGINE: Flat-six
DISPLACEMENT: 2.5L / 152.1 in^3
POWER: 271 hp at 7,900 rpm
TORQUE: 195 lb-ft at 6,200 rpm
GEARBOX: Five-speed
DIMENSIONS (L x W x H): 13' 7.3" x 5' 7.3" x 4' 4.0"
WEIGHT: 2,116 lbs
TOP SPEED: 162 mph

The most successful Porsche representative at the 1972 Monte Carlo Rally, the Gérard Larrousse/Jean-Claude Parramond crew saw victory slip away due to a poor tire choice.

Porsche entered a very special prototype 911 in the 1970 Tour de France Automobile named "psychedelic" due to its decoration. At its wheel, Gérard Larrousse finished third.

911 2.7 CARRERA RS
1972

"Only 500 men will be able to drive it": the first advertisement for the 911 Carrera RS would probably never have been approved today due to its overly divisive signature. Other times, other customs: it was 1972. Deprived of its 5-liter 917 prototypes, banned by legislation at the end of 1971, Porsche reoriented its sports policy towards the new championship reserved for Grand Touring cars. While many considered it to be in its last gasp, the 911 received an unexpected facelift. To comply with the new Group 4 regulations, which, with the exception of wider fenders, prohibited any body modifications, the German manufacturer began work on this special version, to be produced in at least 500 units, considered to be the limit of the demand. Based on the 2.4-liter 911 S, the engineers designed a truly unique model. From its unveiling at the 1972 Paris Motor Show, this new car captivated visitors. The revival of the esteemed "Carrera" designation was not the sole reason for surprise. The Carrera RS sported a ducktail spoiler on the rear hood. Designed by Tilman Brodbeck, this device, working with a fiberglass front spoiler, reduced aerodynamic lift and thus significantly improved road holding at high speeds in competition.

The Carrera RS was the first 911 to be sold with wider wheels at the rear than at the front. It was first fitted with six-inch front wheels and seven-inch rear wheels, all fifteen inches in diameter.

The combination of the light weight and "vitaminized" 911 concept, the 2.7 Carrera RS, underwent an extensive wind tunnel program. The installation of a ducktail-shaped spoiler resulted in a reduction in aerodynamic lift.

ABOVE: Developed for the purpose of homologating a competition version, the 2.7 Carrera RS was officially presented at the Paris Motor Show in October 1972.

RIGHT: Depending on the chosen version, the interior treatment differed. The Sport version was distinguished by the adoption of Glaverbel windows and the removal of the glove compartment, clock, and rear seat.

Given the stress they would have to withstand in races, the suspensions benefited from an improvement program.
The arms guiding the rear wheels were made of light alloy; the suspensions received 15 mm (.59 in) diameter anti-roll bars and Bilstein gas shock absorbers attached to the body by reinforced anchor points.

Against all predictions, the series of 500 units – chassis numbers beginning with 911360 – sold out in a week. Surprised by this

ABOVE: At the beginning of the program, Porsche only planned to produce 500 examples of the 2.7 Carrera RS. Customers had the choice between a Sport (Lightweight) version and a more luxurious Touring finish.

RIGHT: The engineers increased the bore of the 911 24 S engine by 6 mm (.24 in) to 90 mm (3.54 in) to obtain a displacement of 2,687 cm³ and a power identical to that of the 911 R, i.e., 210 hp.

commercial success, Porsche was forced to launch a second series in November 1972. The first examples intended for homologation received special treatment, featuring thinner bodywork, bucket seats, and Glaverbel windows. In addition, Porsche offered an optional Sport or Lightweight version, which, with further weight reduction, dropped the weight to almost 960 kg (2,116 lbs).

The Touring version (M472), which borrowed the interior of the 911 S, weighed 1,075 kg (2,370 lbs). Despite a very high price, production of the 911 Carrera RS reached 1,525 units, including 17 pre-series vehicles and 200 Sport versions (M471 option). Thus, it could be homologated in the Group 3 sports category. Finally, when customers complained that the number of units exceeded 1,500, advertising director Georg A. Ledert produced the clever retort: "Our advertisement only mentioned men. We didn't count our female clientele."

ENGINE: Flat-six
DISPLACEMENT: 2.7L / 164.0 in^3
POWER: 210 hp at 6,300 rpm
TORQUE: 189 lb-ft at 5,100 rpm
GEARBOX: Five-speed
DIMENSIONS (L x W x H): 13' 7.3" x 5' 5.0" x 4' 4.0"
WEIGHT: 2,370 lbs / 2,116 lbs
TOP SPEED: 149 mph

911 2.8 RSR 1973

The Carrera RS served as the basis for the homologation of a competition RSR version, for which the cylinder bore increased by 2 mm (.08 in) to 92 mm (3.62 in), bringing the displacement to 2,808 cm^3. Power reached 296 hp with the adoption of a lighter reciprocating assembly, dual ignition, high-compression pistons, and a vibration damper. At the same time, engineers managed to save more than 80 kg (176 lbs) compared to a Carrera RS with Sport trim by extensively using composite materials for the bodywork and Perspex for the glazing. Meanwhile, the interior ambiance was that of a racing car, featuring two Recaro bucket seats and a roll cage, as well as the removal of all comfort equipment. The widening of the fenders allowed the RSR to be fitted with nine-inch wheels at the front and eleven-inch wheels at the rear. For its participation in endurance events, it was equipped with a 110-liter plastic fuel tank.

The chassis received a battery of modifications. In addition to larger-diameter torsion bars and stiffer shock absorbers, the 917's braking system was enhanced, featuring ventilated and perforated discs, as well as two master cylinders with an adjustable balance bar to vary the brake distribution. Porsche produced only 55 examples of the RSR, primarily for private teams and drivers. The future looked promising for this car as even before its homologation in Group 4, it had won the 1973 24 Hours of Daytona with drivers Peter Gregg and Hurley Haywood.

Porsche produced only 55 examples of the 2.8 RSR, a competition version of the 2.7 Carrera RS. With a body made of composite materials and Perspex glazing, the weight reduction reached 80 kg (176 lbs).

ABOVE: Disguised prototype or improved GT? The 2.8 RSR clearly stands out from the 2.7 Carrera RS from which it is derived. The wider fenders allowed for the installation of 9-inch wheels at the front and 11-inch wheels at the rear.

LEFT: With the 2.8 RSR, Porsche's engineers did not just increase displacement. The adoption of dual ignition and high-compression pistons, among other things, allowed it to reach a power of 296 hp.

ENGINE: Flat-six

DISPLACEMENT: 2.8L / 171.2 in³

POWER: 296 hp at 8,000 rpm

TORQUE: 218 lb-ft at 6,500 rpm

GEARBOX: Five-speed

DIMENSIONS (L x W x H): 13' 7.3" x 5' 5.0" x 4' 4.0"

WEIGHT: 1,852 lbs

TOP SPEED: 177 mph

PORSCHE
S EK 8076

II

TYPE G

—

1973–1989

Against a backdrop of oil crises and economic downturns, the 1970s almost spelled doom for the 911. After complying with new American safety standards by adopting bellows bumpers, and then conforming to noise and pollution control measures at the expense of its essence, it had to face competition from front-engine models. The 911 emerged stronger. In sixteen years, 198,496 units of the G model were produced.

The "Type G" representing the seventh evolution of the 911 since its launch in 1964, logically inherited the seventh letter of the alphabet. Produced from August 1973 – 1974 model year – the new 911 differed significantly from the model it replaced, despite an identical structure. To comply with new safety standards introduced in the United States, its primary market, the 911 was equipped with new bumpers featuring a rubber strip and black plastic shock absorbers at each end. A luminous strip connected the two rear lights. Enhanced safety features also included the installation of three-point seat belts and seats with integrated headrests. New Fuchs wheels in fifteen and sixteen inches with polished rims and five black painted spokes completed the equipment.

The revised six-cylinder boxer engine also evolved. The displacement of all models was increased to 2.7 liters, and although this resulted in a decrease in power, torque increased by 20 Nm (15 lb-ft), providing greater flexibility. As an entry-level model, the 911 (without suffix) with a power output of 148 hp at 5,700 rpm, replaced the 911 T. Succeeding the 911 E, the 911 S delivered 173 hp at 5,800 rpm. The 911 2.7 Carrera retained the characteristics of the RS. The two body variants were maintained: Coupe and Targa. Apart from a few equipment details and the adoption of aluminum-silicon alloy cylinders, the 911 transmission hardly evolved before the 1976 model year. Derived from the Type 916 transmission that equipped the 908 prototypes in the late 1960s, the Type 915 transmission was available with four or five speeds on request. The Sportomatic automatic transmission, still in the catalog, underwent several changes to adapt to the increase in torque, at the cost of reducing the number of gears from four to three. This optional equipment eventually disappeared from the catalog with the 1980 model year.

The period was mainly marked by the appearance of a turbo-powered model. Developed from the Carrera 3.0, Porsche's Supersports car received a 3-liter 256 hp turbo flat-six and a four-speed transmission. Despite the arrival of this model combining luxury and performance, Ernst Fuhrmann, who had presided over the sports brand since the early 1970s, considered that the tightening of American anti-pollution and noise standards would eventually condemn the 911. His judgment was unequivocal: Porsche's flagship model no longer had any room for improvement, especially for competition use. He believed this limitation was inherent to the rear-engine architecture. His convictions were enhanced by the reduction of the 911 S's power to 163 hp for the American market. It was in this context that the first front-engine Porsches appeared: the 924 and the 928.

At the same time, the 911 was content with a few evolutions aimed at improving engine flexibility and efficiency. The increase in bore to 95 mm (3.74 in) increased displacement to 3 liters, but the K-Jetronic injection system resulted in a reduction of 10 hp compared to the 2.7-liter Carrera engine. The 911 Carrera 3.0 thus made do with 197 hp at 6,000 rpm, but the unchanged torque of 255 Nm (188 lb-ft) was achieved at 4,200 rpm instead of 5,100 rpm. This model remained in the catalog for barely more than a year. With the launch of the L series in 1978,

RIGHT: The 911 Turbo, the flagship of the 911 range in the mid-seventies, featured a luxurious finish, including leather upholstery and power windows.

PREVIOUS TWO PAGES: With the appearance of the Turbo version in the mid-1970s, the 911 increased its appeal.

Ernst Fuhrmann defended his strategy by tightening the 911 range. The 2.7-liter and Carrera 3.0 models were replaced by a single model, the 911 SC. Compared to the Carrera 3.0, the flat-six was detuned to 177 hp. A reduction in power, it seemed, to prevent the 911 from overshadowing the 928 and being faster than the brand's new flagship, which was both more powerful and more expensive. With the SC model, the four-speed gearbox was abandoned.

At the same time, Porsche agreed to update the 911 Turbo, which had remained unchanged since its launch. By increasing the displacement to 3.3 liters and installing an air-to-air intercooler between the turbocharger and the engine, the new model produced 296 hp at 5,500 rpm and 411 Nm (303 lb-ft) of torque at 4,000 rpm, resulting in a slightly improved performance. The 3.3-liter could be recognized by its larger rear spoiler, to which a dual exhaust was added from the 1980 model. The increased severity of anti-pollution standards, however, forced the manufacturer to cease exporting the 911 Turbo to the United States and Japan. While this decision vindicated Fuhrmann's policy, it was not to the liking of the Porsche family, which sat on the management board. Disagreements between shareholders and its director became more and more frequent, and during 1980, Fuhrmann was replaced by Peter W. Schutz. The new leader was not an automotive specialist, but, supported by Ferry Porsche, he saved the 911 by allocating it the largest share of the development budget. The power of the 911 SC successively increased to 185 hp, then to 201 hp. Two dormant projects were relaunched: the convertible and the four-wheel drive version. During 1983, the range was expanded with a convertible version developed from the Targa. Porsche 911 enthusiasts could not have asked for more.

Vitality reigned at the Weissach design office, and evolutions followed one another at a frantic pace. The displacement of the flat-six increased to 3.2 liters (3,164 cm^3) with the adoption of the Turbo's 74.4 mm (2.93 in) stroke crankshaft. Power output was 228 hp at 5,900 rpm and torque was 280 Nm (207 lb-ft) at 4,800 rpm. The power of versions intended for markets with the most restrictive anti-pollution regulations was reduced to 204 hp. With the launch of the 911, marketing and sales strategists brought back the Carrera denomination. When the 1987 model year was released (from August 1986), the Type 915 transmission, known for the fragility of its in-house synchronizers and its grinding noises, was replaced by the G50 five-speed gearbox. More pleasant to handle and accepting up to 300 Nm (221 lb-ft) of torque without needing an oil cooler, this new component had been designed to adapt to the 911 Turbo. This materialized in 1989. On the eve of unveiling the third generation of the 911, Porsche reinforced the Turbo's appeal by selling limited Convertible and Targa versions.

911 CARRERA 3.0 RS
1974

The evolution of power and performance eventually relegated the 911 2.8 RSR to a secondary role. This was hardly acceptable for the manufacturer. To regain the top of the rankings, Porsche launched the development of a new sports version, claiming it to be an evolution of the 2.7-liter version, with a planned, limited production of 100 units. Completed in October 1973 and homologated by the FIA on April 1, 1974, the 911 Carrera 3.0 RS pushed the limits of a formula already widely applied by the models that preceded it. While sharing many components with the 2.7 Carrera RS, the new model featured several sports elements that were typically reserved for racing cars. The objective here was to ensure homologation in Group 3 and Group 4 without too much cost. The weight was brought down to around 900 kg (1,984 lbs) with the intensive use of thinner sheet metal and Perspex glazing. The rear spoiler abandoned the ducktail in favor of a horizontal appendage bordered with rubber. The wrap-around bumpers of the 911 Type G and the widened tracks, inspired by those of the 2.8 RSR, allowed the fitting of nine-inch wheels at the front and eleven-inch wheels at the rear. Similar to that of the 2.7 Carrera RS, the interior presentation was simple.

By increasing the cylinder bore to 95 mm (3.74 in), the flat-six's displacement increased to 2,994 cm^3 and power to 230 hp at 6,200 rpm. With this 911/77 block, engineers used a cast aluminum crankcase for the first time, replacing the previously used magnesium. A forged crankshaft resting on plain bearings, forged steel connecting rods, and pistons were also used.

Since Porsche primarily envisioned the car for sports, the Carrera 3.0 RS came with racing brakes similar to those of the 917, an oil cooler in the center of the front apron, and a chassis featuring wider tracks, modified suspension geometry, and large-diameter torsion bars.

The interior of the Carrera 3.0 RS was simple. The steering wheel was that of the standard range models.

Despite being reduced in weight to around 900 kg (1,984 lbs) through the intensive use of thinner sheets and Perspex glazing, the 3.0 RS retained a spare wheel under the front hood.

ENGINE: Flat-six
DISPLACEMENT: 3.0L / 182.7 in^3
POWER: 230 hp at 6,200 rpm
TORQUE: 204 lb-ft at 5,000 rpm
GEARBOX: Five-speed
DIMENSIONS (L x W x H): 13' 10.7" x 5' 9.9" x 4' 4.0"
WEIGHT: 1,984 lbs
TOP SPEED: 155 mph

At Porsche, competition dictated its pace and evolutions. One year after the 2.7 Carrera RS, Porsche launched the Carrera 3.0 RS. This new model, produced in only 55 units, served to homologate a competition version.

911 # CARRERA 3.0 RSR
1974

The 911 Carrera 3.0 RS had no other purpose than to allow the homologation of a competition RSR version. Largely benefiting from the developments of the rolling laboratory that was the Carrera prototype, the Carrera 3.0 RSR thus clearly distinguished itself from its road-going sister. The front and rear fenders were vented. The rear spoiler was lengthened. The fenders, widened by the regulatory 50 mm (1.97 in), accommodated 10.5-inch wheels at the front and 14-inch wheels at the rear. Despite the removal of interior trim and soundproofing, the polyester front hood and lightweight flywheel, the RSR weighed 20 (44 lbs) to 50 kg (110 lbs) more than the road version. This weight gain was because the polyester doors and rear fenders, as well as the titanium mechanical components, were not included in the homologation sheet.

The most significant evolution of the RSR concerned the 3-liter flat-six. Engineers managed to increase power by 100 hp by installing higher lift camshafts, a guillotine intake system, racing connecting rods, and dual breakerless ignition.

Road handling and cornering capabilities also improved significantly. The chassis incorporated the principle used by the second series of the 2.7 Carrera RS and the Carrera 3.0 RS but adapted for competition. At the front, the MacPherson strut uprights were raised by 18 mm (.71 in) to achieve a corresponding lowering of the body. Geometry also evolved, benefiting from negative camber. Equipped with four pistons, the 300 mm (11.8 in) diameter ventilated and perforated discs were lightened.

The work of the engineers was rewarded with brilliant sporting results. During its career, which extended until the end of 1976, the Carrera 3.0 RSR accumulated an impressive series of victories. It won the European Grand Touring Car Championship and the American IMSA Championship in 1974 and 1975. Only 50 units of the Carrera 3.0 RSR were produced.

The 3.0 RSR engine was closer to that of a racing car than that of the Carrera 3.0 RS. The installation of special camshafts, a guillotine intake system, dual ignition, and lightweight racing connecting rods allowed the flat-six to exceed 300 hp.

ABOVE: Preliminary tests for the 1974 24 Hours of Le Mans: the Carrera 3.0 RSRs from Bernard Chenevière's Swiss team and Cachia's French team in the parade.

LEFT: The RSR Carrera body benefited from the lessons learned from the 1974 Carrera RSR Turbo, adopting vents in the front and rear fenders.

ENGINE: Flat-six

DISPLACEMENT: 3.0L / 182.7 in³

POWER: 330 hp at 8,000 rpm

TORQUE: 232 lb-ft at 6,500 rpm

GEARBOX: Five-speed

DIMENSIONS (L x W x H): 13' 10.7" x 5' 9.9" x 4' 4.0"

WEIGHT: 1,984 lbs

TOP SPEED: 177 mph

911

934/935
1976

At Porsche, the production car often served as a homologation base for racing models. This was the case in the early 1970s with the new International Manufacturers Championship regulations. Effective from 1976, it required that Group 4 and Group 5 cars be derived from a homologated production model, produced at a minimum of 400 units. The performance promised by turbocharged engines encouraged Porsche to use this technology. Tasked with defending the colors of the Zuffenhausen manufacturer in the new championship, the 934 (Group 4) and 935 (Group 5) were developed from the 911 Turbo.

Experience gained with the 3-liter 911 RSR, victorious in the 1973 Targa Florio, as well as with the 2.14-liter turbocharged prototype that took second place at the 6 Hours of Watkins Glen and the 24 Hours of Le Mans, also largely contributed to the technical choices. The 3-liter flat-six engine of the 934 featured a boost pressure of 1.4 bar and water cooling of the compressor, allowing for a power output of 478 hp at 7,000 rpm and a torque of 588 Nm (434 lb-ft) at 5,400 rpm. Aside from the reinforced clutch, the addition of an oil pump, and an 80% limited-slip differential, the transmission was similar to that of the 911 Turbo. This was also the case for the bodywork, except for the front and rear bumpers and the widened fenders, allowing the adoption of 15-inch then 16-inch wheels. The very restrictive regulations, especially in terms of weight, exempted it from lightening the body. Apart from the removal of the rear seats and soundproofing, the 934 retained many comfort features, such as electric windows. Entered by private teams, the 934s, produced in only 31 units, dominated the Group 4 category. Meanwhile, the 935 appeared to be much more modified than its sister, with its biplane rear spoiler and very wide rear fenders.

By reducing the bore to 92.8 mm (3.65 in), the engineers reduced the displacement to 2.85 liters. Dual ignition and a six-piston pump ensured a power output of 582 hp at 7,900 rpm. Securing the International Manufacturers Championship title for Porsche in 1976, the 935 continued to evolve in subsequent years. In 1979, a car from the private Kremer team won the 24 Hours of Le Mans.

ENGINE: Flat-six turbo
DISPLACEMENT: 3.0L / 2.9L (182.7 in³ / 174.3 in³)
POWER: 478 hp at 7,000 rpm / 582 hp at 7,900 rpm
TORQUE: 434 lb-ft at 5,400 rpm
GEARBOX: Five-speed
DIMENSIONS (L x W x H): 14' 0.9" x 6' 1.6" x 4' 3.2" / 15' 4.3" x 6' 5.6" x 4' 1.8"
WEIGHT: 2,469 lbs / 2,138 lbs
TOP SPEED: 186 mph

The 934 benefited from an intense program to improve its rigidity involving reinforcements in the front trunk.

RIGHT: From its birth, the 911 Turbo had a sporting extension. To improve its chances in the new Group 4 category, which took effect in 1976, Porsche developed the 934.

BELOW: Tasked with succeeding the Carrera 3.0 RSR, the 934 hardly differed from the 911 Turbo in an attempt to limit costs.

The first 935 model, presented in early 1976, retained the typical 911 fenders. Very quickly, the front end was reworked with the 'flat-nose' principle, with the headlights moved into the bumper.

Unlike the 935, the 934 was never factory-entered. For several seasons, it brought joy to private teams and drivers.

THIER PILS THIER
ADAC
DUNLOP

911 SC SAFARI
1978

Since his participation in the East African Safari in 1973 and 1974, one of the most demanding rallies held in Kenya, Manfred Jantke, the head of the competition department since 1972, dreamt of returning. His ambition materialized in 1978 with the support of the Martini spirits brand, a Porsche partner, and the agreement of Ernst Fuhrmann, the head of Porsche. Two 3-liter 911 SCs were taken from the production line for the German brand's participation in the 1978 Kenyan classic. While they shared similarities with the 2.7 Carrera RS already seen in Africa, particularly in terms of underbody protection and ground clearance, designed to tackle the difficulties of muddy tracks, they were distinguished by several improvements. The rear suspension arms came from the Turbo with special compression and rebound stops. A battery of additional headlights protected by a mesh plate, 280 mm (11.02 in) ground clearance, an engine cover sealed with foam tape to prevent dust from infiltrating the block, special tires mounted on six-inch light alloy wheels gave the 911 SC Safari the perfect adventurer's attire.

With Bosch mechanical injection, the atmospheric 3-liter flat-six prioritized flexibility and reliability. Although detuned, it still delivered nearly 247 hp at 6,800 rpm. With the chosen final drive ratio, the 911 SC Safari could reach nearly 230 km/h (143 mph) when the track allowed. A 40% limited-slip differential and a reinforced clutch completed the setup. Despite extensive equipment including a roll cage, a large fuel tank, and all protective devices, this two-unit vehicle weighed only 1,180 kg (2,601 lbs). While largely dominating the event with drivers Swede Björn Waldegård and Kenyan Vic Preston Jr., the two 911s were delayed by transmission and suspension problems. Although Preston was in second place, the 911s were never seen again on the Safari Rally tracks.

The interior of the 911 SC Safari featured all the equipment necessary for road or long-distance events.

For the 911s entered in the African rallies, the engineers prioritized flexibility and reliability over power. The 3-liter flat-six with mechanical injection thus made do with 247 hp.

ENGINE: Flat-six
DISPLACEMENT: 3.0L / 182.7 in^3
POWER: 247 hp at 6,800 rpm
TORQUE: 286 lb-ft at 5,500 rpm
GEARBOX: Five-speed
DIMENSIONS (L x W x H): 14' 0.9" (width and height unknown)
WEIGHT: 2,601 lbs
TOP SPEED: 143 mph

Raised ground clearance, protective plates, faired underbody, battery of additional headlights mounted on a metal frame, mud flaps behind the wheels, the SC Safari was equipped with the perfect adventurer's arsenal.

911 SC WEISSACH 1979

In 1979, the Porsche commercial brochure presented the 911 SC Weissach as the most exclusive version ever seen in a dealership. Produced in a limited series, the car was destined for the American market. Naming a model "Weissach" is symbolic. Located 25 km from the Stuttgart-Zuffenhausen factory, the village of Weissach has been home to the Porsche Research & Development Center since 1971. It is within this secret location that all the brand's models were born, whether intended for the road or for competition. The 911 SC Weissach thus paid homage to the engineers of the Porsche Motorsport department.

The marketing of this model aimed to support the appeal of the 911 at a time when Ernst Fuhrmann believed that the future of sports cars belonged to front-engine models. History would prove him wrong. Developed from the 911 SC with a 3-liter engine, the Weissach version met American pollution control standards, delivering 173 hp. Intended to showcase all the expertise of the German brand, the special Weissach series benefited from complete equipment designated by option code M439: front spoiler, rear spoiler, passenger mirror, Fuchs wheels with body-colored spokes, fog lights, Bilstein shock absorbers, sport seats, air conditioning system, electric sunroof, and electric antenna. The Weissach was produced in only 468 examples, equally distributed in two colors: black or Platinum metallic (655). The interior was identical in the entire series: beige leather upholstery matched with "burgundy red" carpets with matching piping.

With its versatility of use, the 2+2 architecture significantly contributed to the success of the 911.

ENGINE: Flat-six
DISPLACEMENT: 3.0L / 182.7 in³
POWER: 173 hp at 5,500 rpm
TORQUE: 189 lb-ft at 4,200 rpm
GEARBOX: Five-speed
DIMENSIONS (L x W x H): 14' 1.0" x 5' 5.0" x 4' 4.0"
WEIGHT: 2,745 lbs
TOP SPEED: 140 mph

ABOVE: Produced in a limited series of only 468 examples for the American market, the 911 SC Weissach received the rear spoiler of the famous 911 Turbo and Fuchs wheels painted in the body color.

LEFT: Presented at the time as the most exclusive 911 ever seen in an American dealership, the 911 SC Weissach interior was upholstered in beige leather.

911 SC/RS
1984

Porsche had officially turned away from road events. However, it returned at the beginning of the 1984 season. To meet the new Group B regulations, which defined the premier category of rally cars, Porsche developed the SC/RS. It seems that this project was initiated by the Rothmans tobacco company, which supported the manufacturer in endurance events and wished to maintain a presence in rallying despite the non-renewal of its contract with Opel. The limited investment required – only 20 examples needed to be produced to obtain Group B homologation – largely contributed to the birth of this program. Considered an extension of the 911 SC, the SC/RS nevertheless differed significantly from the production model.

As with all racing models, the SC/RS underwent a slimming diet, reducing its weight to 980 kg (2,161 lbs). The bumpers were made of plastic and fiberglass, while the front and rear hoods, doors, and front fenders were made of aluminum. The window glass was also thin. Inside, soundproofing trim, rear seats, and heating were removed; power windows became manual. The widened body received the running gear of the 911 Turbo, including the brakes and 16-inch tires. The rear spoiler also came from its larger sister.

To comply with competition requirements, the chassis was reinforced with concentric coil springs featuring adjustable threaded stops from 1985 onwards, along with a 40% limited-slip differential.

Derived from the 911 SC, the Type 930/18 3-liter flat-six was distinguished by the installation of a six-piston Kugelfischer mechanical pump instead of K-Jetronic injection, forged pistons, a raised compression ratio, and new camshafts. The power output was 252 hp at 7,000 rpm, but without an air filter, it could reach 280 hp. Meanwhile, the Type 915 gearbox was homologated with several final drive ratios, and the clutch was reinforced.

Despite numerous victories, the 911 SC/RS driven by Finnish driver Henri Toivonen narrowly missed the European Rally Championship title in 1984, finishing second.

The SC/RS model borrowed its dashboard from production models, but, as always with Porsche, the slimming diet involved the removal of numerous comfort features.

ENGINE: Flat-six
DISPLACEMENT: 3.0L / 182.7 in³
POWER: 252 hp at 7,000 rpm
TORQUE: 188 lb-ft at 6,500 rpm
GEARBOX: Five-speed
DIMENSIONS (L x W x H): 13' 10.7" x 5' 9.9" x 4' 2.8"
WEIGHT: 2,330 lbs
TOP SPEED: 149 mph

RIGHT: Despite his talent, Finland's Henri Toivonen missed the European Rally Championship title twice at the wheel of the 911 SC/RS.

BELOW: Born from the new management's decision to relaunch the 911, the SC/RS combined elements of race models and the 911 Turbo.

911 CARRERA 3.2 TURBO LOOK
1984

In parallel with the development of the Carrera 3.2, Porsche explored multiple avenues to maintain the fascination for the 911. The widened body of the 911 Turbo was one of them. Thus, during 1983, Porsche's 1984 flagship model was offered in a Turbo Look version under option code M491. The model did not simply inherit the body of the 911 Turbo; it also borrowed its chassis with larger-diameter anti-roll bars and torsion bars, as well as its larger brake discs and four-piston calipers.

The naturally aspirated flat-six benefited from the latest engineering developments, increasing displacement to 3.2 liters (3,164 cm³) with the 911 Turbo's 74.4 mm (2.93 in) stroke crankshaft, and power to 228 hp at 5,900 rpm, along with torque of 280 Nm (207 lb-ft). At the same time, efficiency was improved by the installation of sequential Bosch Motronic injection.

In 1984 (for the 1985 model year), the 911 Turbo Look could be ordered without the Turbo's rear spoiler, and the range was expanded to include Cabriolet and Targa variants. The Turbo Look model was produced in 3,640 units until 1989. Against all expectations, with 1,684 units, the Cabriolet version was the most requested, ahead of the coupe (1,397 units).

Based on the 911 Carrera 3.2 with the 231 hp flat-six, the Turbo Look version was available as a coupe, convertible, and Targa, the latter being the rarest.

ABOVE: While waiting for the arrival of a new generation of the 911, Porsche maintained interest in the model with a version featuring the bodywork and chassis of the Turbo version.

LEFT: The interior was identical to that of the standard Carrera 3.2 version, but the options catalog allowed customers to personalize their vehicle.

ENGINE: Flat-six

DISPLACEMENT: 3.2L / 193.1 in^3

POWER: 228 hp at 5,900 rpm

TORQUE: 207 lb-ft at 4,800 rpm

GEARBOX: Five-speed

DIMENSIONS (L x W x H): 14' 0.9" x 5' 9.9" x 4' 3.6"

WEIGHT: 2,866 lbs

TOP SPEED: 146 mph

911 CARRERA 3.2 CLUBSPORT 1987

Lighter by 70 kg (154 lbs) compared to a Carrera 3.2, the Clubsport version did not include many comfort features such as rear seats, air conditioning, radio, central locking, and power windows.

During the 1980s, the 911 saw little evolution. Fans of the model despaired at the absence of a "vitaminized" version. Porsche finally acceded to their requests by developing a sport version for the 1987 model year, based on the Carrera 3.2 coupe. By its specifications, this model, named Clubsport (option code M637), was positioned between a standard version and an RS version. As its name indicates, this vehicle was primarily intended for members of Porsche clubs who frequently drove on track.

Compared to the production car, most modifications focused on weight reduction with the removal of the rear seats, air conditioning, fog lights, radio, central locking, and power windows. These elements, combined with the abandonment of much of the sound insulation and the lightening of the body's galvanization system, which reduced the anti-corrosion warranty to two years, contributed to a 70 kg (154 lbs) weight reduction. Lightened by this amount compared to a Carrera 3.2, the Clubsport version dispensed with many comfort features such as rear seats, air conditioning, radio, central locking, and power windows.

At the same time, the engine was optimized, adopting a recalibrated electronic system and new intake valves to push the red line back by almost 600 rpm. While power remained unchanged at 231 hp, it was acknowledged that the best engines could produce up to 250 hp. The revised gearing of the G50 five-speed gearbox improved acceleration and pick-up. The 911 Carrera Clubsport reached 100 km/h (62 mph) 0.5 seconds faster than its production version.

Other modifications included the engine and transmission being mounted on stiffer mounts borrowed from the 911 Cabriolet; Bilstein shock absorbers became standard; the starter was made lighter, and the wiring harness was simplified. Identifiable by its "Carrera CS" red lettering on the lower door panels, this model was produced in only 340 examples.

ENGINE: Flat-six
DISPLACEMENT: 3.2L / 193.1 in^3
POWER: 228 hp at 5,900 rpm
TORQUE: 207 lb-ft at 4,800 rpm
GEARBOX: Five-speed
DIMENSIONS (L x W x H): 14' 0.9" x 5' 5.0" x 4' 4.0"
WEIGHT: 2,646 lbs
TOP SPEED: 152 mph

RIGHT: Between the production Carrera 3.2 model and a sporty RS variant, Porsche created a space for the Clubsport version, which combines comfort with slightly improved performance.

BELOW: Distinguished by "Clubsport" badging on the left front fender and rocker panels, this variant featured a close-ratio gearbox that improved acceleration.

911 SPEEDSTER
1989

During the 1980s, Porsche's management chose to focus its efforts on developing the 911 range. While awaiting the arrival of a new generation, this strategy led to the birth of a stripped-down convertible version, named Speedster, at the Frankfurt Motor Show in October 1987. True to Ferry Porsche's motto that comfort does not create driving pleasure, this model reconnected with the spirit of the 356 Speedster presented in 1954. Porsche hit the mark.

This concept immediately appealed to fans of the German brand. It came at just the right time to contribute to the euphoria of the exceptional car market in Europe and to compensate for the drop in sales in the United States due to the depreciation of the dollar. During the year, Porsche confirmed its decision to produce the 911 Speedster in a limited series. A lucrative business, as this vehicle retailed for more than the 911 Cabriolet, which was more luxurious and better equipped. In the secondary market, some die-hards even paid one and a half times the list price.

Despite very basic comfort, the new Speedster was meticulously developed. From the 911 Carrera Cabriolet, it inherited the 228 hp 3.2-liter engine, the G50 five-speed gearbox, and the chassis. In its final form, it received a manual soft top without lining, requiring serious training before it could be manipulated and stored under the double-hump trunk, and manual side windows without triangular elements. The interior ambiance was that of the Carrera 3.2-liter, except for the seats, the manual heating system, and the removal of the rear seats.

The Speedster offered a truly unique formula around a shortened and more inclined windshield, reduced manual side windows, and a double-hump tonneau cover made of plastic.

ABOVE: Harking back to the 356 Speedster, the concept of a fun car presented at the 1987 Frankfurt Motor Show, based on the Carrera 3.2, led to the birth of the 911 Speedster two years later.

RIGHT: The first 911 Speedster in history was meant to encourage American customers to visit dealerships at a time when the market was starting to decline, and to participate in the European enthusiasm for exceptional models.

The appeal of the car's silhouette lies in its small aluminum windshield, lowered by 80 mm (3.15 in) and showing an inclination five degrees greater than that of the convertible, while with the soft top in place, the glazed area was smaller.

To satisfy customers, two body variants were available from its commercialization for the 1989 model year: Turbo Look and narrow body. Inheriting the widened bodywork and running gear of the 911 Turbo (suspensions, brakes, and wheels, with the exception of torsion bars and stabilizers), the former appeared the most aggressive. This was the preferred of the 2,103 units manufactured. Only 171 units of the narrow-body version were produced, and it is now the most sought-after variant of the series.

ABOVE: The manual emphasizes that using the fully manual hood is only in the event of an emergency. Its operation is a complex procedure that requires some practice.

OPPOSITE: The 1989 Speedster is the last non-series 911 of the G generation. It is available in two body variants: narrow fenders and wide fenders from the Turbo. The latter was favored by customers.

ENGINE: Flat-six
DISPLACEMENT: 3.2L / 193.1 in^3
POWER: 228 hp at 5,900 rpm
TORQUE: 207 lb-ft at 4,800 rpm
GEARBOX: Five-speed
DIMENSIONS (L x W x H): 14' 0.9" x 5' 5.0" x 4' 2.0"
WEIGHT: 2,690 lbs / 2,601 lbs (narrow version)
TOP SPEED: 146 mph

BB·PW 215

III

TYPE 964

–

1988–1994

The 911 celebrated its twenty-fifth anniversary in style with a brand-new generation. The result of fifteen years of development, the 964 generation was the basis of a revolution, introducing all-wheel drive and standard ABS, as well as new comfort features. These advances, however, could not obscure the problems of a company burdened by dependence on the American market, outdated production methods, and the lack of synergy between model ranges.

Was it still possible to improve the 911? After much deliberation and considering its disappearance from the catalog during the 1980s in favor of a range of front-engine sports cars, Porsche answered in the affirmative. As it approached its quarter-century mark, the 911 returned in a totally new version. The directions taken by Peter Schutz materialized at the end of the 1980s. Under the internal designation 964, the third generation was revealed in May 1988. It marked a turning point in the model's history. The German manufacturer's engineers seemed to have found the elixir to keep it up to date. While appearing to be a 911, it consisted of 85% new parts.

Body-colored thermoplastic wrap-around bumpers combined with a retractable rear spoiler integrated into the engine cover achieved the feat of modernizing the silhouette while reducing the aerodynamic drag coefficient (Cx) from 0.39 to 0.32. The spoiler deployed from 80 km/h (50 mph) under the action of an electric motor and returned to its resting position below 15 km/h (9.3 mph). Under the hood was a flat-six engine whose displacement was increased to exactly 3.6 liters by an increase in bore from 95 to 100 mm (3.74 to 3.94 in) and stroke from 74.4 to 76.4 mm (2.93 to 3.01 in). This naturally aspirated Type M64/01 air-cooled engine produced 245 hp at 6,100 rpm and 310 Nm (227 lb-ft) of torque at 4,800 rpm. A first for a production 911: the flat-six used dual ignition (two spark plugs per cylinder), ensuring better fuel combustion and a new electronic management system, Motronic. It was designed to deliver identical power and torque regardless of whether it was equipped with a three-way catalytic converter exhaust, in order to adapt to global air pollution requirements.

Under its modernized bodywork, the 964 boasted transformed road behavior, thanks to the inclusion of a Carrera 4 transmission with all-wheel drive. This development benefited from the experience gained on African tracks with the 953 and the famous 959. The five-speed G50 gearbox distributed part of its power to the front axle via a shaft enclosed in a rigid tube, as on the 959. Unlike the latter, it was equipped with a central differential with epicyclic gear train coupled to an electronically controlled multi-plate hydraulic clutch. The epicyclic train sent 69% of the torque to the rear and 31% to the front. The distribution could reach 100% to the rear or front, depending on wheelspin. The driver could also lock these two clutches for starting on difficult terrain using a control on the center console, but these were released above 40 km/h (25 mph). Completely revised, the running gear adopted an ABS braking system (Anti-lock Braking System), power steering, and a new suspension abandoning torsion bars in favor of coil springs and light alloy arms. Wheel dimensions were identical to those used as standard on the Carrera 3.2, but the famous "Fuchs" wheels were replaced with cast aluminum wheels. The other great innovation was a ventilation system compatible with

RIGHT: The third generation of the 911 used 85% new parts, but kept the instrumentation with its five dials.

PREVIOUS TWO PAGES: A revolution at Porsche. With the 964 generation, presented in May 1988, the 911 was available for the first time with all-wheel drive transmission.

air conditioning. With a mass of 1,440 kg (3,175 lbs), the Carrera 4 reached 100 km/h (62 mph) in 5.9 seconds.

Launched during 1989, the Carrera 4 was reinforced six months later by a Carrera 2 two-wheel drive version, lighter and therefore more powerful. At the same time, a fully automatic four-speed transmission, named Tiptronic, was introduced. Developed in collaboration with Bosch and ZF, it allowed automatic or manual operation, depending on the selector position. In parallel, the Cabriolet and Targa body variants were renewed and completed the Carrera 2 and Carrera 4 range.

As for the 911 Turbo, after a year of absence, it returned to the catalog for the 1991 model year. Developed in barely two years and unveiled at the 1990 Geneva International Motor Show, the new Turbo failed to meet the brand's fans' expectations. It inherited the 964's silhouette and its chassis was borrowed from the Carrera 2, but the evolution seemed too timid to truly convince. It had gained 135 kg (298 lbs), the flat-six engine was still the 3.3-liter from the old generation, and power only increased by 20 hp to 317 hp at 5,750 rpm. The consequent increase in torque to 450 Nm (332 lb-ft) at 4,500 rpm resulted in the disappearance of the brutal acceleration so typical of the first models. Behind these figures, the single-turbo intercooled engine had undergone thorough revision following the adoption of a three-way metallic catalytic exhaust. The compression ratio remained unchanged, but the KKK turbo was replaced with a more efficient model, which operated at a pressure of 0.7 bar, down from 0.8 bar previously. At the same time, the adoption of a dual-mass flywheel and mounting supports reduced vibrations and improved noise filtering. Engine efficiency became more usable. More civilized than its predecessor, the 1990 911 Turbo was equipped with a variable-rate limited-slip differential that engaged at 20% during acceleration and 100% during deceleration. Taking into account the increased performance – 100 km/h (62 mph) was reached in 5 seconds and top speed was 270 km/h (168 mph) – the braking system was resized. Brake disc diameters were increased, from 283 mm (11.14 in) to 298 mm (11.74 in) at the front and from 290 mm (11.42 in) to 299 mm (11.77 in) at the rear. Wheels were increased to 17 inches and the width of the rear tires increased from 225 mm (8.86 in) to 255 mm (10.04 in). Finally, the rack-and-pinion steering was power-assisted. The dashboard featured an airbag steering wheel – a first for Porsche – and a digital computer was integrated into the tachometer. With more direct steering and a hydraulic clutch, the effort required by the driver was less significant than in the past.

Believing that customer complaints had been heard, at the 1992 Paris Motor Show, the Turbo was launched with the 3.6-liter engine. Power gained 40 hp, torque 70 Nm (52 lb-ft). Performance improved significantly.

911 CARRERA CUP 1990

The arrival of the new generation of the 911 prompted Porsche to replace the Turbo Cup single-make championship, raced with 944 Turbo Cups, with the Carrera Cup and 911s. Taken from the Carrera 2 production lines, the Carrera Cup was developed by the competition department of the Weissach center.

The car weighed no more than 1,120 kg (2,469 lbs), a gain of 230 kg (507 lbs) compared to a Carrera 2, with its stripped body and aluminum front hood. The interior equipment was limited to a racing bucket seat with a six-point harness, a welded steel Matter roll cage, and a fire extinguisher. All trims and power windows were removed. The chassis was lowered by 50 mm (1.97 in), the shock absorbers were Bilstein type, and the springs were hardened and shortened.

Carrera Cup drivers could adjust the anti-roll bars to five positions at the front and three at the rear. Toe and camber coefficients were slightly adjustable over a range of three degrees. To avoid excessively rapid tire wear on the rear axle and overheating, Roland Kussmaul and the Weissach team opted for 17-inch five-spoke Speedline wheels mounted on Pirelli slick tires specially designed for the Carrera Cup. The braking system came from the 911 Turbo. In addition to weight reduction, there was an increase in power. The 3.6-liter flat-six gained 15 hp to produce 261 hp at 6,100 rpm.

ENGINE: Flat-six
DISPLACEMENT: 3.6L / 219.7 in³
POWER: 261 hp at 6,100 rpm / 271 hp
TORQUE: 229 lb-ft at 4,800 rpm / 232 lb-ft at 5,000 rpm
GEARBOX: Five-speed
DIMENSIONS (L x W x H): 13' 11.3" x 5' 5.0" x 4' 3.6"
WEIGHT: 2,469 lbs
TOP SPEED: 168 mph

With the appearance of the 911 Carrera RS, Porsche revised the characteristics of the Carrera Cup from the 1992 season. Power increased to 272 hp with the adoption of a modified electronic control unit and more pronounced ignition advance across all rev ranges. The chassis was lowered by 40 mm (1.57 in) and gained a strut brace. The body inherited the lightweight alloy hood and ultra-thin glazing of the RS. The new three-piece detachable aluminum Speedline wheels gained an additional diameter (18 inches). The 1992 Carrera Cup also inherited the Carrera RS braking system. Inside, a button allowed the ABS to be switched off.

To accompany the launch of the new 911 Type 964, Porsche decided from 1990 onwards in Germany to replace the 944 Turbo Cup with the 911 Carrera Cup in its championship iteration.

Produced in 297 units, the Carrera Cup was distinguished from the Carrera 2, from which it was derived, by the removal of comfort features and the reinforcement of the five-speed manual gearbox.

911 964 CARRERA 4 LEICHTBAU 1991

"Why not organize the Carrera Cup around the Carrera 4? That would be an excellent showcase for promoting four-wheel drive technology," said Jürgen Barth, head of the competition department, who in early 1991 launched the development of a car from surplus parts of the Paris-Dakar 953 project. The idea was not to the liking of Ulrich Bez, the technical director, who decided to relaunch the single-make cup around the 964 RS. However, Barth did not abandon his proposal, which led to the birth of the 911 Carrera 4 Lightweight or "Leichtbau" ("lightweight construction").

Not homologated for road use, this model was based on technical achievements and as many widely available Porsche parts as possible. Starting from a bare Carrera 4 body, the weight was kept under 1,098 kg (2,421 lbs) with its aluminum front hood and doors, a carbon fiber rear hood, and fixed side windows featuring sliding Plexiglas panels. The rear spoiler was fixed, the bullet mirrors were attached to the windshield pillars; the exhaust was free; an oil filler flap was installed on the right rear fender, and the interior was stripped down with only two bucket seats and a bolted-in roll cage. The 3.6-liter flat-six delivered 265 hp, and the gearbox came from the Type 953 Paris-Dakar. The ultra-short ratios favored acceleration at the expense of a top speed of only 210 km/h (130 mph). A reinforced competition clutch and a single-mass inertia volume completed the equipment. The driver could adjust the torque distribution between the four wheels via two rotary controls on the center console.

About twenty enthusiasts, mainly from American and Japanese Porsche clubs, embraced this concept, which did not aim for optimal efficiency but provided all the sensations of a basic racing car.

Initially designed for the Carrera Cup, the lightweight, four-wheel-drive 964 eventually gave rise to a small series of about twenty cars not homologated for road use.

ABOVE: Unlike the Carrera 4, the lightweight Leichtbau version adopted a fixed rear spoiler from the Turbo and fixed side windows with a sliding panel.

RIGHT: The driver could adjust the torque distribution between the four wheels via two rotary controls on the center console.

ENGINE: Flat-six

DISPLACEMENT: 3,600 cm³

POWER: 265 hp at 6,720 rpm

TORQUE: 304 Nm at 5,250 rpm

GEARBOX: Five-speed

DIMENSIONS (L x W x H): 14' 0.3" x 5' 5" x 4' 3.9"

WEIGHT: 1,089 kg (approx. 2,421 lbs)

TOP SPEED: 260 km/h (approx. 162 mph)

911 TURBO IMSA SUPERCAR 1991

To participate in the American "IMSA Bridgestone Potenza Supercar Championship" series, established in 1991, Porsche developed a competition version of the 911 Turbo, the model that offered the best guarantees of victory. Regulations allowed for few modifications compared to the production model, aiming to avoid cost escalation and prioritize spectacle. The engine retained its 3.3-liter displacement and was pushed to the limits allowed by regulations with a boost in pressure, with power reaching 377 hp at 6,000 rpm. Most improvements focused on the adoption of a second oil cooler and better cooling of the flat-six. The gearbox was standard, but the clutch was reinforced. The endurance of the braking system relied on competition linings. The chassis benefited from a 40 mm (1.57 in) lowered suspension, stiffer shock absorber springs, and larger-diameter anti-roll bars. The fuel tank volume was reduced to 60.5 liters, and a welded roll cage reinforced both the passenger compartment in the event of an impact and the car's overall rigidity. The regulations imposed a 100 kg (220 lbs) ballast, bringing the weight of the 911 Turbo IMSA to 1,429 kg (3,150 lbs), despite the use of a bare chassis.

Known for its extensive experience in motorsport, the Brumos Racing team was entrusted with operating the car. At the end of the 1991 season, American Hurley Haywood won the championship.

Porsche defended the title the following year, and supported throughout the season by German Hans-Joachim Stuck and rally driver Walter Röhrl in the final rounds, the German manufacturer claimed another crown, ahead of Lotus. The 1993 season confirmed the 911 Turbo's dominance. By winning seven of the nine events and taking pole position in every race, Stuck again won the title, but without Haywood, who competed in a sister car.

Racing under the banner of the American Brumos team, the 911 Turbo captured the IMSA Supercar Championship for Grand Touring cars three consecutive years.

Beneath the typical silhouette of the 911 Turbo, the IMSA Supercar version had a reinforced braking system and clutch.

The 911 Turbos entered in the American IMSA Supercar Championship differed little from the production model. While the bodywork was identical, the engine benefited from a few modifications.

ENGINE: Flat-six turbo

DISPLACEMENT: 3.2L / 195 in³

POWER: 380 hp at 6,000 rpm

TORQUE: 361 lb-ft at 4,800 rpm

GEARBOX: Five-speed

DIMENSIONS (L x W x H): 14' 0.3" x 5' 10" x 4' 2.8"

WEIGHT: 3,150 lbs

TOP SPEED: 177 mph

911 964 CARRERA RS
1992

The aluminum front hood, space-saver spare wheel, reduced electrical wiring, and anti-corrosion protection contributed to reducing the weight of the Carrera RS by 160 kg (353 lbs) compared to the Carrera 2.

From the beginning of 964 production, Porsche began investigating a sports version based on the Carrera 2. This new model pursued a dual objective: to serve as a homologation base for a racing car intended for the N/GT (National Grand Touring) category and to help reposition the brand upward. Presented at the Paris Motor Show in September 1990, this new 911 reconnected with the best of the brand's history, resurrecting the "Carrera RS" (RennSport) designation, eighteen years after the 2.7 Carrera RS.

Enthusiasts flocked to the car, drawn by its technical specifications, which closely resembled those of a Carrera Cup version, and its limited production. They would not be disappointed by this new RS, which relied on a proven recipe: a weight loss and vitamin treatment. The RS was recognizable by its "Carrera RS" rocker panel stripes, its Turbo-type mirrors, and its wheels shared with the Carrera Cup.

A strict two-seater, the 911 Carrera RS was characterized by the removal of numerous comfort features, including sound insulation and the rear bench seat.

Engineers succeeded in reducing the weight to 1,220 kg (2,690 lbs). A saving of around 160 kg (353 lbs) was achieved by adopting an aluminum front hood and magnesium wheels, as well as by reducing electrical wiring and anti-corrosion protection. The extremely stripped-down interior was distinguished by the exclusion of key comfort features, including rear seats, soundproofing, central door locking, electric mirrors and windows, automatic heating control, defrosting, air conditioning, power steering, a hi-fi system, an alarm system, and a dome light. The RS was equipped with lightweight leather Recaro bucket seats that matched the body color, and a door opening system utilizing a fabric strap. The 3.6-liter flat-six now bore the designation M64/03.

ABOVE: The return of the Carrera RS to the catalog, a designation laden with symbolism, was part of the strategy to regain customers at a time when Porsche was going through a difficult period.

LEFT: The Carrera RS's performance was significantly superior to that of the Carrera 2. It reached 100 km/h (62 mph) in 5.3 seconds, with a top speed of 260 km/h (162 mph).

Due to its 40 mm (1.57 in) lowered chassis and stiffer suspensions, the 911 Carrera RS has always been characterized by a certain lack of comfort on degraded asphalt roads.

ENGINE: Flat-six

DISPLACEMENT: 3.6L / 220 in^3

POWER: 260 hp at 6,100 rpm

TORQUE: 240 lb-ft at 4,800 rpm

GEARBOX: Five-speed

DIMENSIONS (L x W x H): 14' 0.3" x 5' 5.0" x 4' 2.4"

WEIGHT: 2,703 lbs

TOP SPEED: 162 mph

The lightweight door panels, strap door openers, and bucket seats make the Carrera RS interior close to a standard 911.

With a simple modification in the electronic control unit and special camshafts, this dual-ignition unit gained 10 hp to produce 257 hp at 6,100 rpm. The increased torque delivered 325 Nm (240 lb-ft) at 4,000 rpm. Derived from the Carrera 2, the five-speed gearbox featured closer ratios for the first two gears, enabling 100 km/h (62 mph) to be reached in 5.3 seconds, 0.4 seconds faster than the Carrera 2. The chassis also evolved with a lowered ride height of 40 mm (1.57 in), stiffer springs, and anti-roll bars with three adjustment positions. The braking system was in line with the performance: large ventilated and perforated discs with four-piston calipers from the 911 Turbo at the front 322 mm (12.68 in) and from the Carrera Cup at the rear 299 mm (11.77 in). For N/GT homologation, the Carrera RS was equipped with 205/50 ZR 17 tires at the front and 255/40 ZR 17 at the rear.

In addition to the 1992 examples produced for Europe, there were 290 lightweight N/GT versions, stripped of all comfort features and equipped with a roll cage, as well as 701 units of the American version. Its homologation for the American market required the adoption of an airbag steering wheel and power windows. A sport suspension, a fixed Turbo rear spoiler, and a detuned engine producing 247 hp.

The interior of the Turbo S Leichtbau was inspired by that of the Carrera RS, with manual windows, leather-upholstered Recaro bucket seats, and fabric straps for door opening.

911 TURBO S LEICHTBAU 1992

The mood was not celebratory at Porsche in the early 1990s. In 1991, production declined to fewer than 21,000 units, despite a still well-developed range structured around four model families: 911, 944 S2, 968, and 928. Customers shunned the front-engine models, and the 911 ensured the company's survival. All avenues were explored to increase turnover and support the brand's image. In this context, Rolf Sprenger, the director of Porsche Exclusive, suggested exploiting the 911 Turbo's victories in the IMSA Championship with a special version treated in the same spirit as the 911 Carrera RS.

Named 911 Turbo S Leichtbau, this car was presented at the 1992 Geneva Motor Show. Noting the enthusiastic reception it received, Porsche approved the project in May to produce a small run for 295,000 Deutsch Marks. Considered the technological showcase of the Weissach design office, this super Turbo almost achieved the performance of the celebrated 959 by combining weight reduction and increased power. Engineers succeeded in lowering the weight to 1,290 kg (2,844 lbs), a gain of 180 kg (397 lbs) compared to a production 911 Turbo, by using carbon-fiber-reinforced plastic composite for the front and rear hoods, doors, and rear spoiler, as well as by removing numerous comfort features.

The 3.3-liter turbocharged flat-six engine power increased to 381 hp at 6,000 rpm and its torque reached 490 Nm (361 lb-ft) at 4,800 rpm. Performance took a serious leap forward. The 911 Turbo S could reach 100 km/h (62 mph) in 4.6 seconds and a top speed of 292 km/h (179 mph).

These advances led to revising the tire recommendations in favor of a more generous setup, with 235/40 ZR 18 tires at the front and 265/35 ZR 18 tires at the rear. The chassis was completely revised. It was lowered and stiffer springs, shock absorbers, and anti-roll bars installed. Only 86 units of the 911 Turbo S Leichtbau were produced.

Developed from a 911 Turbo by applying the same modifications as a Carrera RS, the 911 Turbo S Leichtbau's weight lightened by 180 kg (397 lbs) and its power increased by 51 hp.

ENGINE: Flat-six turbo
DISPLACEMENT: 3.3L / 201 in^3
POWER: 381 hp at 6,000 rpm
TORQUE: 361 lb-ft at 4,800 rpm
GEARBOX: Five-speed
DIMENSIONS (L x W x H): 13' 11.3" x 5' 9.9" x 4' 2.8"
WEIGHT: 2,855 lbs
TOP SPEED: 179 mph

Presented at the Geneva Motor Show in March 1992, the 911 Turbo S Leichtbau was considered the technological showcase of the German manufacturer. Its production was limited to 86 units.

911 SPEEDSTER 1992

In survival mode at the beginning of the 1990s, due to falling sales leading to abyssal losses, Porsche summoned the best of its past to maintain its desirability. While deliveries of the Carrera RS had barely begun, the manufacturer unveiled a new Speedster at the 1992 Paris Motor Show. Despite a lower price than the Carrera 2 Cabriolet from which it was derived, this new member of the 911 family did not achieve the expected success. Only 936 examples out of the 3,000 planned were produced. The fiasco stemmed from the absence of a wide-fender Turbo Look version, like its predecessor. Produced on the same assembly line as the Carrera 2 and Carrera 4, the Speedster was recognizable by its lowered windshield, 17-inch Cup-design wheels, and its double-hump rear trunk in body color. It was only available in four colors: Speed Yellow, Maritime Blue, Grand Prix White, and Indian Red. The Speedster concept was preserved: the soft top was still manual, but its operation was considerably revised and improved compared to the 1989 model.

Powered by the naturally aspirated 3.6-liter flat-six, it delivered 250 hp. It was mated to the five-speed manual gearbox or the Tiptronic automatic transmission, upon request. The car was delivered to some markets with a 40% limited-slip differential as standard. The two-seater interior dispensed with the main comfort features of a Carrera 2. Air conditioning and the onboard computer were optional features, while the strap-operated door opening system and leather-upholstered bucket seats were derived from the Carrera RS. Standard or sport seats were, however, available on request. Ultimately, the Speedster spirit was verified by a 50 kg (110 lbs) weight saving, bringing the total weight down to 1,350 kg (2,976 lbs).

During this model's career, Porsche eventually acceded to certain customers' requests by producing 15 examples on a wider Turbo body. 469 units would also be destined for the North American market.

Like the RS models, the 964 Speedster version featured a lightweight door opening system.

Appearing at the end of the 964 generation's career, the Speedster was developed in record time and reused many elements from the Carrera 2 dashboard.

At Porsche, when times turn difficult, history is often called upon. Thus, the 964 eventually appeared in the Speedster silhouette, though initially only with narrow fenders. Presented at the 1992 Paris Motor Show and developed from the 964 Carrera 2 Cabriolet, the model failed to achieve the expected success.

ENGINE: Flat-six

DISPLACEMENT: 3.6L / 220 in³

POWER: 250 hp at 6,100 rpm

TORQUE: 229 lb-ft at 4,800 rpm

GEARBOX: Five-speed

DIMENSIONS (L x W x H): 13' 11.3" x 5' 5.0" x 4' 2.0"

WEIGHT: 2,976 lbs

TOP SPEED: 162 mph

911 CARRERA RS 3.8
1993

In parallel with the competition version Carrera 3.8 RSR, the Weissach competition department developed a road version called the Carrera RS 3.8. It represented the pinnacle of lightweight two-wheel-drive 911s. The front hood, as well as the doors, were aluminum, similar to the Carrera RS, while the rear hood was made of carbon-Kevlar. Like the RSR, the side and rear windows were made of extra-thin glass. Cooling scoops replaced the fog lights in the front bumper. From the RSR, the civilian version inherited the widened body of the 911 Turbo 3.6 and the enormous plastic biplane rear spoiler was adjustable to six positions.

The interior was as stripped down as that of the Carrera RS. The seats were only adjustable with a wrench. The weight did not exceed 1,210 kg (2,668 lbs). Like the competition model, the flat-six displacement increased to 3,746 cm^3, but in a detuned version producing 297 hp at 6,500 rpm. Torque reached 360 Nm (266 lb-ft) at 5,250 rpm. The gearbox, identical to that of the Carrera RS, operated with a reinforced clutch mechanism. The limited-slip differential was no longer variable but fixed at 40%. The axle geometry originated from the 911 Carrera, not the Turbo. The lowered and stiffened suspension had adjustable anti-roll bars, with five positions at the front and three at the rear. A spacer above the 92-liter fuel tank stabilized the front suspension, especially during emergency braking. The Bilstein gas shock absorbers, with specific damping, had a spring rate 50 to 70% firmer. Engineers focused primarily on improving comfort, which was not the strong point of the Carrera RS.

The 911 Carrera 3.8 RS is undoubtedly the most captivating version of the 964 generation. Walter Röhrl would not say otherwise!

BB-PW 256

ABOVE: The silhouette of the 911 Carrera 3.8 RS is very close to that of the Turbo S Leichtbau, and only the fenders and rear spoiler allow them to be differentiated.

Although it did not fully utilize the potential of its wider tracks, the RS 3.8 was equipped with Dunlop Sport wheels measuring 235/40 ZR 18 at the front and 285/35 ZR 18 at the rear. Brake disc diameters were increased: 342 mm (13.46 in) at the front and 315 mm (12.4 in) at the rear. Performance took a leap forward: top speed was 270 km/h (168 mph) and 100 km/h (62 mph) was reached in 5 seconds. Sold at a price of 225,000 DM when a Carrera 3.6 cost 80,000 DM, the RS 3.8 was produced in only 55 units. It is now one of the most prized models of the 964 generation.

BELOW: At Porsche, nothing is lost. The interior of the 911 Carrera 3.8 RS was largely inspired by that of the Turbo S Leichtbau with its Recaro bucket seats upholstered in leather and the removal of the rear seats.

OPPOSITE: The 911 Carrera 3.8 RS is easily recognizable, mainly by its large plastic biplane rear spoiler, which can be adjusted in six positions.

ENGINE: Flat-six
DISPLACEMENT: 3.7L / 229 in^3
POWER: 300 hp at 6,500 rpm
TORQUE: 284 lb-ft at 5,250 rpm
GEARBOX: Five-speed
DIMENSIONS (L x W x H): 14' 0.3" x 5' 9.9" x 4' 2.8"
WEIGHT: 2,668 lbs
TOP SPEED: 168 mph

RS 3,8

911 CARRERA 3.8 RSR
1993

After setbacks in Formula 1 and Indy Car, Porsche showed keen interest in the revival of Grand Touring. This return to its roots allowed the company to demonstrate, once again, the speed and endurance qualities of its models, which were directly derived from series production.

During 1992, in anticipation of the various GT championships about to be introduced in Europe and the United States, Porsche announced the development of a new racing car. The engineering team, led by Roland Kussmaul, embarked on a race against time to be ready by early 1993. Named the Carrera 3.8 RSR, Porsche's new GT weapon was developed around a lightweight 911 Turbo 3.6 shell, featuring an aluminum front hood and doors, and a steel rear hood that supported the large composite spoiler of a Carrera RS. The flat-six engine displacement increased to 3,746 cm^3, with a bore of 102 mm (4.02 in). The new catalyzed M64/04 block produced 325 hp, but the examples entered in competition approached 380 hp. It differed from the 3.6-liter with individual cylinder fueling via six separate throttles, more elaborate Bosch Motronic 2.10 management, as well as shorter and lighter pistons, special valves, and tappets. A double oil cooler, a 120-liter fuel tank, a roll cage, a pneumatic lift, and center-lock wheels completed the equipment. The weight did not exceed 1,135 kg (2,502 lbs) thanks to a completely stripped interior with a single racing bucket seat.

In total, Porsche Motorsport produced 51 Carrera 3.8 RSRs, including the 49 commercially available cars, the development prototype, and a factory race car. Workshop records indicate that four additional chassis were produced in the autumn of 1994.

While it retained the dashboard of the RS road version, the Carrera RSR dispensed with all comfort features, retaining only a bucket seat and safety harnesses.

ABOVE: Roll cage, pneumatic lift, center-lock wheels, biplane spoiler topped with an additional device: the RSR was equipped with the full panoply of a true racing car.

RIGHT: With a flat-six engine increased to 229 in³ and approaching 380 hp, the 911 Carrera 3.8 RSR almost reached 186 mph top speed on the fastest circuits.

ENGINE: Flat-six

DISPLACEMENT: 3.7L / 229 in³

POWER: 325 hp at 6,900 rpm / 365 hp at 7,400 rpm

TORQUE: N/A

GEARBOX: Five-speed

DIMENSIONS (L x W x H): 14' 0.3" x 5' 9.9" x 4' 2.4"

WEIGHT: 2,425 lbs

TOP SPEED: 179 mph

911 TURBO S LE MANS 1993

Of the eleven 911 cars lined up at the start of the 1993 24 Hours of Le Mans, the white 911 number 46 driven by Walter Röhrl, Hans-Joachim Stuck, and Hurley Haywood caught the attention of connoisseurs. This factory-entered model, supervised by Norbert Singer, differed in many respects from other competition 911s.

When the Fédération Internationale de l'Automobile (FIA) announced at the end of 1992 that endurance events would be open to Grand Touring cars, sports car manufacturers accelerated their project development. Porsche revealed its cards in the form of a spectacular 911 Turbo in February 1993, at the Paul Ricard circuit in Provence. Ostensibly recalling the famous silhouettes of the mid-1970s – 934 and 935 – and due to its imposing biplane rear spoiler and enormous wheels (27/65 X18 at the front and 30/70 X18 at the rear), the 911 Turbo S LM-GT was more of a prototype than a GT.

Starting from a completely stripped-down 911 Turbo S, the competition department engineers managed to reduce the weight to 1,050 kg (2,315 lbs) by utilizing a Kevlar body reinforced with carbon in specific areas. The engine came from the 962 prototype, which belonged to the Group C category. This 3.2-liter unit, featuring two spark plugs and two valves per cylinder, was entirely air-cooled and supercharged by two KKK turbochargers, delivering a power output of between 480 and 500 hp.

While the Provençal test drive hinted at interesting potential, nothing replaced the race itself. The factory entered the new car in the 12 Hours of Sebring, where the three companions – who would later be found at Le Mans – finished seventh overall and first in the GT category. At Le Mans, the 911 S LM-GT retired at nightfall following an accident. Sold to a brand collector, the car continued its career in early 1994 with a 3.6-liter engine, under the banner of the French Larbre Compétition team. At the 24 Hours of Daytona, Dominique Dupuy, Jesús Pareja, Bob Wollek, and Jürgen Barth drove it to second place overall and first in GT. Other victories would follow in the nascent BPR Championship (Barth - Peter - Ratel) events.

OPPOSITE: Unlike the Carrera RSR, the 911 Turbo S Le Mans is more of a prototype than an improved version of a Grand Tourer.

ABOVE After competing in the 12 Hours of Sebring and the 24 Hours of Le Mans under factory colors, the Turbo S Le Mans continued its career under the banner of a private French team.

ENGINE: Flat-six turbo
DISPLACEMENT: 3.2L / 193 in³
POWER: 500 hp at 7,500 rpm
TORQUE: 361 lb-ft at 5,400 rpm
GEARBOX: Five-speed
DIMENSIONS (L x W x H): N/A
WEIGHT: 2,315 lbs
TOP SPEED: 186 mph

911 30 JAHRE 1993

At Porsche, history takes on a sacred dimension. Thus, as the 911 approached its thirtieth anniversary, the manufacturer conceived a special version. This new model was not only to celebrate three decades of success, but also to close the chapter on the 964 generation.

Porsche had already announced the launch of a completely new car. Unveiled in March 1993, this anniversary model was based on the all-wheel-drive Carrera 4 coupe and featured a 247 hp 3.6-liter flat-six engine. To encourage sales, the manufacturer equipped it with the most desirable bodywork, the extra-wide Turbo Look, and announced a limited series production, numbered to 911 examples. All vehicles were recognizable by the "911 30 Jahre" (911 30 Years) badge on the rear engine cover and the numbered plaque on the rear shelf. Only three metallic body colors were available – violet, amethyst, and polar gray – but most vehicles would be delivered in the first of these. Identical to the Carrera 4, the chassis nevertheless featured specific suspension arms. As always at Porsche, the few privileged individuals who ordered the 964 30 Jahre could personalize it by drawing from the extensive options catalog. Given the success of this limited series, Porsche would produce 164 additional vehicles identical to the anniversary car, but without any specific reference.

ABOVE: In the early 1990s, the first steering wheels equipped with an airbag were not a paragon of elegance.

RIGHT: For the 911's thirtieth anniversary, Porsche returned to a tried and true formula: a 911 Carrera with the chassis and bodywork of a 911 Turbo.

To celebrate thirty years of the 911, Porsche preferred the all-wheel-drive chassis of the Carrera 4 over the two-wheel-drive Carrera 2.

ENGINE: Flat-six
DISPLACEMENT: 3.6L / 220 in^3
POWER: 250 hp at 6,100 rpm
TORQUE: 229 lb-ft at 4,800 rpm
GEARBOX: Five-speed
DIMENSIONS (L x W x H): 13' 11.3" x 5' 10.1" x 4' 2.8"
WEIGHT: 3,417 lbs
TOP SPEED: 158 mph

911 TURBO 3.6 "FLACHBAU" 1993

The 911 Turbo S X85 models boasted the ultimate in comfort equipment offered by Porsche in the mid-1990s.

A 911 Turbo with lowered front fenders and retractable headlights was nothing new. From the late 1970s, the factory's special department produced 911 Turbos on demand with this flat-nose front hood, resembling the 935 racing version. The most famous was the one ordered by Mansour Ojjeh, the owner of TAG Group Limited and sponsor of the Porsche V6 Turbo engine for the McLaren team, in the early 1980s. During the same decade, the Stuttgart-Zuffenhausen firm continued to produce customized 911 Turbos, whether coupes or even convertibles.

Porsche revived the formula during 1993, but the operation was on a completely different scale from that of the 1970s and 1980s. It was no longer about individually produced cars, but a series of 76 coupes that elegantly marked the end of 3.6-liter 911 Turbo (Type 964) production before the introduction of the 993 generation. Porsche Exclusive was tasked with the transformation. The cars had lowered fenders and retractable 928-style headlights at the front. The front bumper integrated additional air intakes for the oil cooler. The rear spoiler featured painted edges, while the wheels were identical to those of the 911 Turbo S. The features extended beyond the bodywork. The Type M64/50S X88 engine was derived from that of the competition 911 Turbo, which produced 382 hp at 5,750 rpm. It was mated to a five-speed manual gearbox and a 40% ZF differential. Half of the production of these limited-edition 911 Turbos was delivered to the United States. Ten examples went to Japan.

The 911 Turbo S X85 flat nose models were technically close to the 911 Turbo S Leichtbau.

ENGINE: Flat-six turbo
DISPLACEMENT: 3.6L / 220 in³
POWER: 385 hp at 5,750 rpm
TORQUE: 384 lb-ft at 5,200 rpm
GEARBOX: Five-speed
DIMENSIONS (L x W x H): 14' 0.3" x 5' 9.9" x 4' 2.8"
WEIGHT: 3,241 lbs
TOP SPEED: 181 mph

Inspired by the famous 935 engaged in competition, the "flat nose" front with painted fenders gave rise to a special series based on the 964.

S·LW 7858

IV

TYPE 993

1993–1998

Under the advertising slogan: "You think you know it, but you won't recognize it," the fourth generation of the 911 carried profound modifications. The aluminum chassis, multi-link rear axle, and the adoption of twin turbochargers for the Turbo version were the main innovations. A transitional model introduced during Porsche's extreme financial difficulties, the 993 remained on the market for barely more than four years. It was the last 911 to be equipped with an air-cooled engine. Its production did not exceed 68,881 examples.

Neither quite the same, nor quite different, the 911 passed its thirtieth anniversary with a host of improvements. The 964 had not even had a five-year run, to the detriment of the company's profitability. Known internally by code number 993, the fourth generation made its first public appearance at the Frankfurt Motor Show in September 1993. The extent of its evolution was not apparent from the windshield, roof, and doors retained from the prior version. Designer Harm Lagaay's team successfully modernized the timeless silhouette of the sports car without compromising its original form. The now-integrated bumpers highlighted the harmony of the whole. The silhouette was more streamlined than on previous models, thanks to the lowered front fenders in the style of the 959 and the abandonment of the round headlight in favor of projector-style headlights reminiscent of the 968. Flush windows and a new double-sweep wiper placement ensured a wider field of vision and better wiping quality. The more bulbous front hood increased trunk capacity by 20% to 123 liters. The generous retractable rear spoiler and faired underbody contributed to airflow and reduced aerodynamic drag. The new self-supporting body, made of galvanized steel with a welded box section, increased torsional rigidity by 20% compared to the 964. Besides its looks, the 993 improved in handling, gearbox precision and flexibility, comfort, and sound insulation.

It was a revolution: instead of the oblique arm and elastic leg axle used until then, engineers installed a multi-link rear axle with an LSA (Lightness, Stability, and Agility) system. Made of cast aluminum, the auxiliary subframe of the rear axle was connected to the body at four points via rubber bushings, improving comfort and handling. It was combined with a revised front axle – featuring increased caster, a negative scrub radius, and offset anchor points – and more generous tire sizes (16 and 17 inches, optional). The result was remarkable agility and enhanced steering precision. Porsche's reputation in braking was strengthened by the adoption of larger diameter discs, four-piston calipers, and standard ABS 5 anti-lock braking system.

The latter was combined, another first, with an ABD (Automatic Brake Differential) traction control system reducing torque to the slipping wheel. 911 enthusiasts rejoiced. The flat-six engine retained its characteristic melody. Still air-cooled and with a displacement of 3.6 liters, it developed 263 hp at 6,100 rpm and 330 Nm (243 lb-ft) of torque at 5,000 rpm, thanks to a lighter reciprocating assembly. Electronic management for ignition and injection increased engine efficiency while reducing fuel consumption. Hydraulic valve lash adjustment, a first, combined with a new two-part exhaust system in the widened rear fenders and large volume metallic catalytic converters reduced pollution. The new 911 Carrera already met the noise and air pollution reduction standards of 1996.

Another innovation contributing to economy and comfort was a new six-speed manual gearbox that perfectly exploited the engine's increased power. Still available, the four-speed Tiptronic

RIGHT: The dashboard of the fourth-generation 911 differs little from that of its predecessor.

PREVIOUS TWO PAGES: Presented in September 1993, the 993 was the last generation to feature an air-cooled engine.

gearbox was equipped with steering wheel controls. The 993 also focused on onboard comfort with a more luxurious finish, redesigned control ergonomics, two standard airbags, and a more efficient air conditioning system.

1994 was marked by the introduction of Cabriolet versions with a fully automatic wind deflector and the Carrera 4. The all-wheel drive system of the latter was reduced in weight by 50 kg (110 lbs) and equipped with the ABD system. The model lineup was expanded in autumn 1995 with the introduction of a Targa variant. It introduced a patented concept based on an electrically operated glass roof that retracted behind the rear window. The main distinguishing feature of the Targa was bolted two-piece wheels. With the 1996 model year (from August 1995), the 3.6-liter flat-six engine increased to 286 hp, utilizing the variable intake system called "Variocam".

The paradox of the 993 was that it appealed beyond the fan base, but the financial crisis of the early 1990s reduced demand, and Japanese competition and high production costs harmed its profitability. And as the front-engine models disappeared from the catalog, Porsche was threatened with collapse. The management multiplied initiatives to strengthen the appeal of the 911. This is how the Carrera S and Carrera 4S variants were born, featuring the bodywork and chassis of the new 911 Turbo. The latter made a splash at the Geneva Motor Show in March 1995. This time, German engineers developed a model that avoided criticism from an increasingly demanding clientele. Based on the 3.6-liter flat-six, the Turbo became the most powerful production Porsche in history. With its two KKK turbochargers, a first, it delivered 397 hp at 5,750 rpm and 540 Nm (398 lb-ft) at 4,500 rpm. From 2,500 rpm, torque reached 450 Nm (332 lb-ft), utilizing airflow regulation via a bypass valve and Bosch Motronic M5.2 electronic management.

Already affecting Porsche's technical developments, environmental concerns were reflected in the flagship of the range with a continuous air quality monitoring system. The 911 Turbo was equipped with a new six-speed gearbox and an all-wheel-drive system adopting the operating principle of the 911 Carrera 4. The 993's multi-link rear axle chassis was also retained with modifications, allowing full exploitation of the increased performance. The wheel arches were widened by 25 mm (.98 in) at the rear, a fixed spoiler distinguished the rear end, and wheel dimensions were increased to 235/40 ZR 18 at the front and 285/30 ZR 18 at the rear. Top speed culminated at 290 km/h (180 mph) and 100 km/h (62 mph) was reached in 4.5 seconds. Braking efficiency relied on a disc diameter increased to 322 mm (12.68 in), up from 299 mm (11.77 in) previously, and a five-channel ABS system. Finally, the 911 Turbo introduced hollow-spoke aluminum wheels. This design technique saved 11 kg (24 lbs).

911 CUP 3.8
1995

When transitioning from the 964 to the 993 for the Carrera Cup single-make race series, the Motorsport department drew inspiration from developments made on the latest competition models. Following a well-proven formula, Roland Kussmaul and his team designed a 993 Carrera Cup representing significant evolution. The team in charge of the project managed to reduce the weight by 20 kg (44 lbs) compared to the 964 generation, bringing it down to 1,100 kg (2,425 lbs). The small quarter windows and the rear window were made of Plexiglas, and the side windows were made of thin, lightweight glass. The front hood was aluminum, the doors lacked side impact protection, and the interior was completely stripped.

Like the production model, the Carrera Cup engine was developed under the direction of engineer Herbert Ampferer. The Type M64/70 3.8-liter flat-six gained a resonance intake system, allowing it to produce 310 hp at 6,100 rpm, with a maximum torque of 360 Nm (266 lb-ft) at 5,500 rpm. The maximum permissible engine speed was increased to 6,900 rpm, despite the valve train from the 3.8 RSR being capable of 7,800 rpm. These evolutions would not have been possible without the reinforcement of the six-speed transmission with steel synchronizer rings.

On the chassis side, the 993's multi-link rear axle allowed for higher performance. The Cup 3.8 benefited from a 70 mm (2.76 in) reduction in ride height compared to a Carrera 2, as well as new shock absorbers and springs that are twice as stiff as those of the previous model. Tire manufacturer Pirelli developed new tires, 235/45 X18 at the front and 285/45 X18 at the rear. The five-channel ABS braking system was equipped with larger ventilated and drilled brake discs and four-piston calipers. Welded to the shell, the FIA-certified roll cage contributed to the car's rigidity. Developed and perfected during 1993, the 993 Carrera Cup began its sporting career the following year in the Supercup International series and entered the various national series in the 1995 season. On this occasion, the car gained the fixed rear spoiler developed for the Carrera RS Clubsport and new chassis settings. From 1996, the power of the naturally aspirated flat-six was increased to 315 hp at 6,100 rpm and torque to 370 Nm (273 lb-ft) at 5,500 rpm.

In parallel with the development of the production 993, the competition department produced a Carrera Cup version that began its sporting career in 1994.

ENGINE: Flat-six
DISPLACEMENT: 3.7L / 229 in^3
POWER: 310 hp at 6,100 rpm / later 315 hp at 6,100 rpm
TORQUE: 266 lb-ft at 5,500 rpm / 273 lb-ft
GEARBOX: Six-speed manual
DIMENSIONS (L x W x H): 13' 11.1" x 5' 8.3" x 4' 2.0"
WEIGHT: 2,425 lbs
TOP SPEED: 174 mph

ABOVE: During its career, the 993 Carrera Cup evolved into three versions. In total, 186 vehicles were manufactured for the single-make series, resulting in consistently spectacular races.

RIGHT: At Porsche, they prefer evolution to revolution. Motorsport is no exception, and the 993 Carrera Cup version focused on improving existing components.

911 CARRERA RS
1995

Porsche seemed to have found the key to endlessly prolonging the 911's appeal. At the Amsterdam Motor Show in January 1995, it unveiled a new Carrera RS version based on the 911 Carrera Type 993. Like previous models in the RS lineage, the new model combined lightweight construction with a power revision. Starting from the 911 Carrera 2, based on the principle that lightness aids road holding, engineers hunted down superfluous kilos. The 911 Carrera RS achieved a 100 kg (220 lbs) weight reduction to 1,270 kg (2,800 lbs). This slimming cure would not have been possible without sacrificing comfort features. Thus, electric window controls, central locking, headlight washers, intermittent wiper control, and the radio pre-equipment kit were removed.

This intensive hunt led to reducing the windshield washer fluid reservoir capacity from 6.5 liters to 1.2 liters, and the sophisticated Carrera lighting system was replaced by the small bulbs of the 911 Speedster. Driver and passenger airbags were available on request. The use of thin glass and the omission of rear window defrosting further saved 5 kg (11 lbs). The aluminum hood alone represented a saving of 7.5 kg (16.5 lbs). The enveloping bucket seats with shells painted in body color saved 30 kg (66 lbs). The interior was devoid of soundproofing materials, and the doors opened with a fabric strap.

BELOW: Equipped with the new Variocam variable valve timing system, based on variable-length air intake manifolds, the Carrera RS's flat-six engine delivers 300 hp.

RIGHT: Don't be fooled by appearances. On board the Carrera RS, the ambiance evokes the production model, but the performance and driving sensations are unique.

ABOVE: Derived from the world of competition, the 3.8-liter flat-six of the 993 Carrera RS's maximum engine speed increased to 6,900 rpm.

RIGHT: The 993 Carrera RS was fitted with the new six-speed manual gearbox, but the first three gears were lengthened compared to the production model.

Porsche sold two variants of the Carrera RS: a Clubsport version, recognizable by its biplane rear spoiler, and a production version close to a Touring finish.

At the same time, the dual-ignition, dry-sump flat-six displacement increased to 3.8 liters (3,744 cm³) by increasing the bore from 100 to 102 mm (3.94 to 4.02 in). With its modified valves, camshafts, and valve train, as well as the adoption of the new Variocam variable valve timing system based on fluctuating intake manifold lengths, the Carrera RS announced 300 hp at 6,500 rpm and a torque of 355 (262 lb-ft) at 5,400 rpm. Power was transmitted to the road via a six-speed manual gearbox and 18-inch wheels. The first three gears were lengthened to coincide with the speed range achieved on the track. A dual-mass flywheel was added to reduce transmission noise. The chassis was lowered by 30 mm (1.18 in) at the front and 40 mm (1.57 in) at the rear, and the anti-roll bars had five adjustment steps at the front and three at the rear.

The Carrera RS was equipped as standard with the ABD (Automatic Brake Differential) and a mechanical limited-slip differential for the rear axle. The kinematics of the axles were modified for track use. Stiffer steering bearings led to more precise wheel guidance. A double lip under the front bumper, side skirts, Cup-design wheels, and the rear spoiler completed the standard equipment. A Clubsport variant was produced with an imposing spoiler, a roll cage, specific bucket seats and harnesses, a single-mass flywheel, and a shorter fifth gear.

Unveiled at the 1995 Amsterdam Motor Show, the 993 Carrera RS was distinguished from the Carrera 2 by its unique front and rear spoilers.

ENGINE: Flat-six

DISPLACEMENT: 3.7L / 229 in^3

POWER: 300 hp at 6,500 rpm

TORQUE: 262 lb-ft at 5,400 rpm

GEARBOX: Six-speed manual

DIMENSIONS (L x W x H): 14' 0.3" x 5' 8.3" x 4' 2.0"

WEIGHT: 2,800 lbs

TOP SPEED: 171 mph

ENGINE: Flat-six biturbo
DISPLACEMENT: 3.6L / 220 in³
POWER: 430 hp at 5,750 rpm / 450 hp
TORQUE: 398 lb-ft at 4,500 rpm
GEARBOX: Six-speed manual
DIMENSIONS (L x W x H): 13' 11.1" x 6' 1.2" x 4' 1.8"
WEIGHT: 2,789 lbs / 2,535 lbs
TOP SPEED: 183 mph

911 GT2 1995

The euphoria that the launch of the 911 Carrera RS and the 911 Turbo created among fans of the German brand reached new heights with the GT2. Brutish: the most fitting description for this customer competition model presented at the Geneva Motor Show in 1995. With its riveted fiberglass fender flares and monumental biplane rear spoiler with air intakes at the ends, Porsche's new competition department flagship was recognizable at first glance.

Based on the two-wheel-drive 911 Turbo, the GT2's genesis was guided by an uncompromising motto: to deliver unfiltered speed sensations. Roland Kussmaul, the project manager, applied a proven Porsche recipe: shedding superfluous kilos and increasing power. The GT2 weighed only 1,290 kg (2,844 lbs), compared to 1,500 kg (3,307 lbs) for a production 911 Turbo, by using aluminum doors and hood lids, ultra-thin side and rear windows, a rear window without defrosting, and by reducing onboard equipment.

The 3.6-liter twin-turbo flat-six (Type M64/60) was boosted to 430 hp at 5,750 rpm using a boost pressure of 0.9 bar and Motronic M5.2 electronic management. Torque reached new heights: 540 Nm (398 lb-ft) at 4,500 rpm. A sign of great flexibility, it never dropped below 450 (332 lb-ft) between 2,000 and 4,000 rpm. The GT2 managed to eclipse the famous 959 by boasting a better power-to-weight ratio. Equipped with a six-speed manual gearbox, the 911 GT2 accelerated from 0 to 100 km/h (62 mph) in 4.3 seconds and passed 200 km/h (124 mph) in 12.5 seconds. Top speed was set at 295 km/h (183 mph). The 30 mm (1.18 in) wider fenders accommodated three-piece Speedline wheels with magnesium cross-spokes, 235/40 ZR 18 at the front and 295/35 ZR 18 at the rear. The chassis, lowered by 20 mm (.79 in) compared to the 911 Turbo, featured a three-level adjustable rear anti-roll bar, a special auxiliary subframe mounting, a 40/60 limited-slip differential, and racing-derived spring-dampers. Aerodynamics were not forgotten: the rear spoiler was adjustable and the front bumper lips resembled those of the Carrera RS Clubsport.

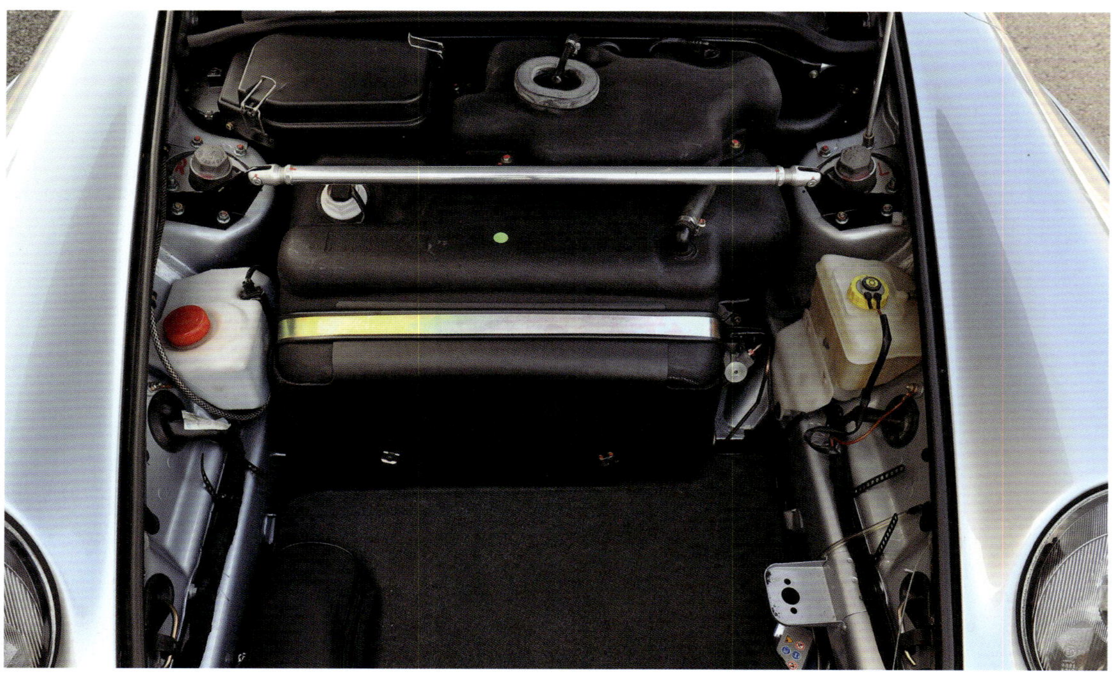

Developed for homologation of a competition version, the road-going 911 GT2 borrowed many technical solutions from the world of racing.

Upon its revelation in 1995, the 911 GT2 appeared as the most impressive model since the introduction of the 911 in 1963.

Porsche 911 enthusiasts immediately recognized the 993 GT2 with its air intakes at the ends of its biplane rear spoiler.

911 GT2 COMPĒTITION
1995

On the cover of the January 1995 *Christophorus* magazine, sent to the brand's customers, the 911 GT2 race car set the tone. With this car, Porsche continued the grand tradition of GT competition cars. Intended for customers who wanted to participate in endurance and sprint races around the world, this new model succeeded the 911 Carrera RSR 3.8. Based on the 911 Turbo, the GT2 was distinguished by its riveted plastic fender flares and a biplane rear spoiler with side air intakes for the turbochargers. It borrowed most of its solutions from the Carrera RS Clubsport and the Carrera RSR 3.8. The weight was reduced to 1,150 kg (2,535 lbs) by utilizing an aluminum front hood and doors, as well as thin glass for the rear and side windows. The interior layout was identical to that of the Carrera RSR.

The power-to-weight ratio was 2.55 kg (5.6 lbs)/hp with the 3.6-liter boxer flat-six's power, increased to 450 hp. Hidden under an enormous intercooler, the engine incorporated competition-specification Motronic management and two KKK turbochargers, blowing at 0.9 bar, positioned to the left and right of the cylinder heads.

The bodywork, widened by 32 mm (1.26 in) at the front and 30 mm (1.18 in) at the rear, was set on 18-inch three-piece BBS (Baumgartner Brand Schiltach) wheels with center locks, and widths of 10 and 11 inches at the front and rear respectively. The suspension was almost identical to that of the production Turbo, but it received additional Unibal spherical joints and, at the front, a strut brace and a five-level adjustable anti-roll bar. The rear axle was equipped with specific springs and shock absorbers and a three-level adjustable anti-roll bar.

The escalation of performance led Porsche to develop a GT2 Evolution. Presented at the Frankfurt Motor Show in 1995 with a harlequin livery, the new GT2 was homologated in the GT1 category. It was distinguished by its modified front spoiler with additional air intakes for brake and oil cooler ventilation, as well as its larger rear spoiler. Power was increased to 600 hp.

The weight of the GT2 Compétition was reduced to 1,150 kg (2,535 lbs) by using aluminum body components and thin glass windows.

ENGINE: Flat-six biturbo
DISPLACEMENT: 3.6L / 220 in³
POWER: 450 to 485 hp at 5,750 rpm
TORQUE: N/A
GEARBOX: Six-speed manual
DIMENSIONS (L x W x H): 13' 11.1" x 6' 1.2" x 4' 1.8"
WEIGHT: 2,425 lbs
TOP SPEED: 186 mph

The increase in performance led the competition department to improve the GT2's performance. By the end of its career, it approached 600 hp.

Presented with a harlequin livery at the Frankfurt Motor Show in 1995, the 911 GT2 Compétition was homologated in the GT1 category.

911 GT1
1996

Is it still a GT? The question is not misplaced. Apart from its front fenders and rear lights, the 911 GT1 bore a resemblance to a prototype. Since Grand Touring cars were given a major role in endurance racing in the mid-1990s, Porsche had demonstrated its expertise with the GT2. However, faced with the escalation of car performance and the entry into the GT1 category, the German manufacturer found itself at a loss. The response came in the form of the 911 GT1, developed under the direction of Herbert Ampferer, competition director, and Norbert Singer, project director. The first prototype covered its first laps on the Weissach test track on March 14, 1996.

Entered at Le Mans in June, Porsche's new weapon preceded the commercialization of the road car by one year. To cut short all rumors, the manufacturer nevertheless took care to bring a registered GT1 to Sarthe. While the entire front part of the chassis reused the underbody of the 911, the powertrain and the rear suspension layout clearly referred to the Porsche 956 and 962 that dominated Group C in the 1980s. However, the technical study was entirely new and specific to the GT1. Inspired by studies conducted on Group C cars, the water-cooled flat-six with two KKK turbos was installed in a mid-rear position. With a displacement of 3,164 cm^3, it developed 600 hp at 7,200 rpm, accompanied by a phenomenal torque of 650 Nm (479 lb-ft) at 5,500 rpm. Instead of having three separate cylinder heads as in the 962 engine, each cylinder bank was topped with a single cylinder head. The four camshafts were driven by two chains, rather than by gears. Under the carbon hood, the famous flat-six engine of the production version produced 544 hp at 7,000 rpm, weighing 1,075 kg (2,370 lbs). The six-cylinder was mated to a six-speed gearbox and a single-plate clutch.

The bodywork illustrated the excesses of the GT championship in the mid-1990s. A true prototype, the 911 GT1 borrowed its engine from Group C prototypes of the 1980s.

The interior of the cars participating in the 24 Hours of Le Mans had nothing in common with that of a production 911.

ENGINE: Flat-six biturbo

DISPLACEMENT: 3.2L / 193 in³

POWER: 544 hp at 7,000 rpm

TORQUE: 443 lb-ft at 4,250 rpm

GEARBOX: Six-speed manual

DIMENSIONS (L x W x H): 15' 5.4" x 6' 4.8" x 3' 10.2"

WEIGHT: 2,370 lbs

TOP SPEED: 190 mph

To homologate its 911 GT1 for competition, Porsche had to build a small series of road cars.

The GT1 excelled in the field of aerodynamics, with downforce never seen before on a 911 race car. Even more than during acceleration, this efficiency was particularly evident in braking, with deceleration values exceeding 1g above 100 km/h (62 mph). Despite replacing the carbon discs of the race cars with steel brakes of the same diameter, the 911 GT1 went from 0 to 200 km/h (124 mph) in just 10.4 seconds before stopping in 130 meters, according to Weissach engineers.

In the road-homologated version, Porsche recreated the atmosphere of the 911's interior, but the width of the side skirt, the lowered ride height, and the reinforcing tube that crossed the clear opening when the door was opened, hindered access to the interior. Space was limited and overheated despite the air conditioning.

The 1997 version differed significantly. The more enveloping front hood accommodated the headlights of the new Boxster and 911 Type 996. The front tracks were widened by 10 cm, the suspension was redesigned, and the mechanics were enhanced. At Le Mans, victory eluded the official GT1s. The first retired due to a run-off, the second due to fire. It took a third participation at the Le Mans circuit to favor the 911 GT1. In its fiftieth year, Porsche achieved a one-two finish with a significantly modified version of its flagship vehicle. The highly refined bodywork rested on a carbon fiber monocoque chassis and housed a twin-turbo flat-six engine whose displacement was increased to 3,196 cm³. Assembled individually by the Motorsport department located within the Weissach Research & Development Center, 36 examples of the GT1 were produced.

The 911 GT1 straddles two generations of 911. The version introduced in 1997 anticipated the headlights of the new 997 generation.

A special series designed by the Porsche Exclusive department, the 991 Turbo S brings the 993 generation to a beautiful close.

911 TURBO S
1997

While the 911 Carrera Type 996 had already been unveiled, Porsche decided to conclude the 911 Turbo's career in style. It entrusted its "Personalization" department, now renamed "Porsche Exclusive," with designing a special series. Affixed with the suffix "S," this special version was produced in only 345 units, of which 183 were for the American market.

Unlike the production Turbo, the S version was enhanced with an aerodynamic kit that included a specific front bumper and a double-blade rear spoiler. The chassis was lowered by 15 mm (.59 in) and the brake calipers were painted yellow. Their cooling was reinforced by air ducts located near the fog lights in the front apron, as well as by air intakes on the rear fenders.

From the legendary 959, the 911 Turbo S inherited the electronically controlled all-wheel drive system. Two double exhaust pipes located at each end of the rear bumper and polished 18-inch wheels contributed to differentiating the model.

The interior ambiance, more refined than that of a production 911 Turbo, featured full leather upholstery, white-faced instrumentation, and carbon trim for the dashboard, door panels, center console, tunnel, and gear lever. To accompany these evolutions, the 3.6-liter twin-turbo flat-six was revisited. It now produced 450 hp due to modified electronic management. The gearing of the six-speed gearbox was also revised. Result: the Turbo S accelerated from 0 to 100 km/h (62 mph) in 3.6 seconds.

ENGINE: Flat-six biturbo
DISPLACEMENT: 3.6L / 220 in^3
POWER: 450 hp at 6,000 rpm
TORQUE: 427 lb-ft at 4,500 rpm
GEARBOX: Six-speed manual
DIMENSIONS (L x W x H): 13' 11.1" x 6' 1.0" x 4' 1.8"
WEIGHT: 3,307 lbs
TOP SPEED: 183 mph

OPPOSITE: Mainly sold in the United States, the special Turbo S series was distinguished by white-faced instrumentation. Every detail mattered. The rear carpet featured the Turbo S signature.

turbo S

S GO 1097

V

TYPE 996

1997–2005

The fifth generation of the 911 marked a real turning point in its history, ushering in water-cooled flat-six engine technology. Its career took off just as the brand's founder, Ferry Porsche, passed away on March 27, 1998. Designed on a common platform with the Boxster, the 996 allowed the Zuffenhausen brand to return to profitability. Its career was distinguished by a profusion of variants and production volumes never before achieved.

In Stuttgart, smiles returned to faces. Appointed head of Porsche in 1993, Wendelin Wiedeking intensified the program of production rationalization and productivity improvement, as well as opening up to new markets. He approved the new strategy, sacrificing the less popular front-engine models and refocusing around the 911 and a second completely new model that still aligned with the brand's DNA. In the minds of Porsche's managers, the pursuit of economies of scale necessitated that the two vehicles share a maximum number of technical elements. They were therefore developed simultaneously, but for purely economic and commercial reasons, the German manufacturer gave priority to its first truly new car to come out of the factories in eighteen years. Named Boxster (Type 986) like the 1993 Detroit Motor Show concept, this model took the form of a sporty roadster whose lines evoked the 550 Spyder of the 1950s. Presented at the 1996 Paris Motor Show, the car shared 40% of its parts with the 911 Carrera Generation 996, including the front bodywork, steering gear, dashboard, and engine base. The fenders and hood were perfectly interchangeable. When the fifth generation was presented at the 1997 Frankfurt Motor Show, the public became aware of the many similarities between the two Zuffenhausen sports cars.

Produced until 2005, the fifth generation of the 911 represented the most radical evolution of the model since its introduction in 1963. Built on an entirely new platform, the car gained an additional 185 mm (7.28 in) in length, with 120 mm (4.72 in) of this increase benefiting the wheelbase. This resulted in better straight-line stability and more generous interior space. While the styling became more subdued compared to the 993, the bodywork was more voluminous – the front track measured 50 mm (1.97 in) more – and 50 kg (110 lbs) lighter with the use of very light but 50% stiffer HLE (High Elastic Limit) steel. The reinterpreted classic design was rewarded with a drag coefficient reduced to 0.30 instead of 0.34.

Under the rear overhang, there was a flat-six engine derived from the Boxster's, but rotated 180 degrees to sit behind the axle. Despite a decrease in displacement compared to the 993, power was up. Reduced to 3.4 liters but with its four-valve technology, the new flat-six already produced 297 hp at 6,800 rpm and 350 Nm (258 lb-ft) of torque at 4,600 rpm. Average consumption decreased by 10% and thanks to variable valve timing and intake, the torque had astonishing elasticity. The 996 was the first 911 to feature a water-cooled engine. This technology was chosen to ensure optimal cooling of the four valves per cylinder, improve efficiency, and also comply with tightening anti-pollution standards worldwide. The Variocam system for camshaft timing based on engine speed, later optimized as Variocam Plus, provided significant advantages. At low revs, the variation in valve timing reduced polluting emissions while stabilizing the idle speed. At medium revs, the system employed a timing that increased torque.

Available over a wide rev range, this engine excelled in acceleration, taking 5 seconds less than the 993 to go

RIGHT: Designed on a common platform with the Boxster, the fifth generation of the 911 marked a true turning point in the model's history.

PREVIOUS TWO PAGES: Introducing a new 34-liter water-cooled flat-six, the fifth generation appeared in autumn 1997.

from 100 (62 mph) to 200 km/h (124 mph) in fifth gear. Top speed was increased to 280 km/h (174 mph) and 100 km/h (62 mph) was reached in 5.2 seconds. The perfected chassis, which could receive a Sport version, prioritized driving comfort and quiet operation. The interior was completely revised. The ambiance was more refined and the dashboard shared its center console with the Boxster. It used the Porsche Communication Management (PCM), an integrated and innovative communication system. The change took the form of a redesigned instrument cluster with five intertwined dials, but the tachometer remained in the center. The equipment was enhanced with an adjustable steering column, redesigned seats, electric windows and mirrors, and four standard airbags. The 996 introduced the PSM (Porsche Stability Management) stability control as an option.

Shortly after the launch of the 911 Carrera, the Cabriolet and Carrera 4 returned to the catalog. The all-wheel drive version continued to evolve. A maximum of 40% of the torque could be transmitted to the front axle via the viscous coupling now housed in the front differential housing. The PSM stability program contributed to the vehicle's traction. The Carrera 4 was available for the first time with the Tiptronic S five-speed transmission. At a dizzying pace, 911 models and variants multiplied. At the 1999 Frankfurt Motor Show, it was the new 911 Turbo's turn to appear. Derived from the 3.6-liter dry-sump GT3 engine, the flat-six produced 416 hp at 6,000 rpm and 560 Nm (413 lb-ft) of torque at 2,700 rpm. The Variocam Plus system could be paired, for the first time, with the Tiptronic S gearbox. Performance reached a new level, with a 4.2-second acceleration to 100 km/h (62 mph) and a top speed of 305 km/h (190 mph). With rear fenders widened by 65 mm (2.57 in) compared to a 911 Carrera, the Turbo received a sport chassis lowered by 10 mm (.39 in) and a reinforced braking system.

The following year, the 911 Carrera underwent a facelift. While the redesigned front gained personality with the Turbo's headlights, the flat-six was converted to the 3.6-liter with Variocam Plus. It produced 320 hp at 6,800 rpm, for a torque of 370 Nm (273 lb-ft) at 4,250 rpm. The 911 Carrera benefited from these evolutions to reach 285 km/h (177 mph) and cross 100 km/h (62 mph) in 5 seconds. The appeal of the 911 range was strengthened with the appearance, a few months apart, of a Targa version and a Carrera 4S with coupe or convertible bodywork, distinguished by rear fenders widened by 60 mm (2.36 in), 18-inch wheels, reinforced braking, and rear lights connected by a luminous strip. Porsche concluded the 996 chapter in style by reintroducing the 911 Turbo Cabriolet from the 2004 model year.

911 GT3
1999

In the interest of consistency with its competition counterpart, the new sport version of the 911 Carrera abandoned the Carrera RS designation in favor of GT3. Presented at the Geneva Motor Show in March 1999, this sport variant adopted a 3.6-liter flat-six derived from the 911 GT1 victorious at Le Mans in 1998, with a power of 360 hp at 7,200 rpm and a torque of 370 Nm (273 lb-ft) at 5,000 rpm.

Even before its commercialization, the GT3 earned its stripes as a reference in sports cars. At its wheel, rally world champion Walter Röhrl completed the 20.8 km Nürburgring circuit in Germany in 7 min 56 sec. A performance based on revised chassis linkages: lowered chassis by 30 mm (1.18 in), adjustable anti-roll bars and stiffer shock absorbers, extended adjustment range for axle geometries, reinforced braking system with large diameter discs and four-piston fixed calipers, reinforced monobloc aluminum type. The interior lost nearly 30 kg (66 lbs) with the removal of the rear seats and the use of leather sport bucket seats.

Porsche underestimated the potential of the GT3, and the 500 examples were sold so quickly that a second batch of 1,350 units was built. In early 2003, for the 911's fortieth anniversary, a new evolution of the GT3 emerged. The manufacturer immediately approved a production of 2,500 units over one year for this model, commercialized for the first time in the United States. The flat-six gained 15 hp and 15 lb-ft thanks to a reduction in moving masses and the introduction of the Variocam system. With a power per liter of 105.8, this 3.6-liter became one of the best naturally aspirated engines in the world. The GT3 also served as the basis for the Cup version launched in 1998. Paired with a six-speed manual gearbox, the 3.6-liter flat-six with a power of 355 hp. During its career, the Cup would undergo mechanical and aerodynamic evolutions. From the 911 GT3 Cup, the competition department developed a GT3 R version intended for endurance events, which was then further developed into a GT3 RS version with two air intake restrictors. The lineage ended in 2004 with the GT3 RSR entered in the GT2 category.

The 991 GT3 interior lost nearly 30 kg (66 lbs) with the removal of the rear seats and the installation of leather sport bucket seats.

ENGINE: Flat-six
DISPLACEMENT: 3.6L / 220 in³
POWER: 355 hp at 7,200 rpm / 376 hp at 7,400 rpm
TORQUE: 273 lb-ft at 5,000 rpm / 284 lb-ft
GEARBOX: Six-speed manual
DIMENSIONS (L x W x H): 14' 6.4" x 5' 9.5" x 4' 2.0"
WEIGHT: 2,976 lbs / 3,042 lbs
TOP SPEED: 188 mph / 190 mph

RIGHT: In 1999, the GT3 succeeded the RS sports car. The new 3.6-liter flat-six, now water-cooled and with four valves per cylinder, produced 360 hp and 370 lb-ft of torque.

BELOW: The 911 GT3 underwent an evolution during the 911's fortieth anniversary. The 3.6-liter flat-six was now capable of revving to 8,200 rpm, a gain of 400 rpm, and the rear hood incorporated a larger spoiler.

911 CUP / GT3 R
1999

At Porsche, everything is intertwined. The sport versions of road models and competition versions are always very similar. Thus, with the 996 generation, the GT3 and its racing derivatives were developed in parallel. Succeeding the 964 Cup, from 1998, for the European Supercup single-make series, and from the following year, in the various national Carrera Cup championships, the 996 Cup was considered the competition version of the GT3. To reduce costs, only essential modifications were made. Around a bare chassis, the competition department focused on increasing the 3.6-liter flat-six's power to 380 hp and optimizing the suspension for slick tires.

In another part of the workshop, the GT3 and Cup models gave birth to the GT3-R customer competition version, developed to participate in all GT races worldwide. Before its commercialization at the end of 1999, Porsche conducted a full-scale test by assigning two cars to the 24 Hours of Le Mans with private teams. The experience proved positive as the two experimental 996 GT3s finished the Sarthe event, with the fastest in thirteenth place. These two new-generation 911s were equipped with a 3.6-liter engine featuring four water-cooled valves per cylinder.

With the GT-R, engine builders increased power to 430 hp and the red line to 9,000 rpm. Specific shock absorber spring settings and the adoption of multi-level adjustable anti-roll bars distinguished the chassis. The body was equipped with synthetic fender flares and an enormous rear spoiler. The competition department produced 60 examples of the 996 GT-R. The 996 Cup also enjoyed great success, with Porsche producing 873 units across its various versions.

BELOW & MIDDLE: Developed in parallel with the GT3 and Carrera Cup, the 996 GT3 R largely benefited from competition lessons. Its bodywork gained wider fenders and a massive rear spoiler.

ABOVE: Like most modern race cars, the 996 GT3 R was equipped with an electronic tachometer and displayed a multitude of information on a digital screen. The steering wheel was detachable.

ENGINE: Flat-six
DISPLACEMENT: 3.6L / 220 in³
POWER: 377 hp at 7,000 rpm / 387 hp at 7,300 rpm / 427 hp at 8,200 rpm
TORQUE: 288 lb-ft at 6,300 rpm / 288 lb-ft at 6,350 rpm
GEARBOX: Six-speed manual
DIMENSIONS (L x W x H): 14' 6.4" x 5' 9.5" x 4' 2.0"
WEIGHT: 2,579 lbs / 2,469 lbs
TOP SPEED: + 186 mph

Considered the queen of endurance in its category, the 996 GT3 R was distinguished by its low fuel consumption and optimized efficiency.

911 MILLENNIUM 2000

The German manufacturer never missed an opportunity to celebrate an event. For the turn of the new millennium in 2000, Porsche marketed a commemorative version of the 911 coupe based on the 996 Carrera 4.

Named Millennium and produced in only 911 numbered examples worldwide, this special model received a dark purple (Purple Pearl Metallic) paint with reflections that changed with the light. This color was combined with 18-inch wheels and chromed oval exhaust tips, as well as a polished 911 logo on the engine cover. The latter housed a revised version of the 3.4-liter M96 flat-six, whose power had been increased to 300 hp.

The interior was entirely covered in caramel-colored leather. Very refined, the 911 Millennium featured carpets coordinated with the upholstery and instrumentation with white-faced dials. In addition to the sunroof, the Millennium model's equipment included electrically adjustable seats, cruise control, a numbered plaque at the bottom of the center console, a suede headliner, the 911 logo on the door sills, and maple burl inserts for the steering wheel, gear lever, and handbrake.

ABOVE: Based on the four-wheel drive 911 Carrera 4, the Millennium version was produced in 911 numbered examples worldwide.

ENGINE: Flat-six
DISPLACEMENT: 3.4L / 207 in³
POWER: 300 hp at 6,800 rpm
TORQUE: 258 lb-ft at 4,600 rpm
GEARBOX: Six-speed manual
DIMENSIONS (L x W x H): 14' 6.4" x 5' 9.5" x 4' 3.4"
WEIGHT: 3,031 lbs
TOP SPEED: 174 mph

RIGHT: Thanks to the 996, and also to its little sister, the Boxster, with which it shared many parts, Porsche entered the 21st century reinvigorated.

OPPOSITE: Distinctive ambiance for the 911 Millennium, celebrating the arrival of the new millennium. The instrumentation featured a white background and the interior was entirely covered in caramel-colored leather.

911 GT2
2001

With a view to homologation for competition in the GT category, Porsche relaunched a sport variant of the 911 Turbo. Marketed in early 2001, the new 911 GT2 was clearly inspired by the GT3. The most powerful 911 in the range was immediately identifiable by its open-worked bumper with four air intakes and an aerodynamic lip, its yellow-painted brake calipers, its imposing rear spoiler, and its monstrous 18-inch wheels shod with extra-wide tires (315/30 ZR 18 at the rear).

The GT2 inherited the 3.6-liter twin-turbo flat-six, but a boost increased pressure to 2 bar, new intercoolers, and variable valve lift and camshaft timing (Variocam Plus), so that it produced 462 hp at 5,700 rpm and a constant torque of 620 Nm (457 lb-ft) from 3,500 to 4,500 rpm. Dry sump lubrication with a separate oil reservoir met the specific requirements for intensive track use. Similarly, the six-speed gearbox was reinforced.

In parallel with the increase in power, the GT2 shed 100 kg (220 lbs) at 1,440 kg (3,175 lbs) with the removal of the rear seats and all-wheel drive. Designed to be very sporty, this version was equipped with a 20 mm (.77 in) lowered chassis, adjustable anti-roll bars, stiffer suspension compatible with racing springs, a limited-slip differential, and standard ceramic brake discs.

Thus equipped, the GT2 shattered all records. Not content with reaching 100 km/h (62 mph) in 4.1 seconds and pushing speeds up to 315 km/h (196 mph), it only needed 7 minutes 47 seconds to complete a lap of the Nürburgring Nordschleife. These sprinter performances did not prevent a meticulous finish.

ABOVE: The 2001 911 GT2 did not settle for a vitamin and slimming cure. The attention paid to aerodynamics included a front lip and air intakes in the bumper and rear fenders.

LEFT & OPPOSITE: The dashboard was almost identical to that of a 911 Turbo, but the focus on weight reduction meant removing the rear seats, as well as the radio and air conditioning, which were available on request. With a boost pressure of 2 bar, the 3.6-liter turbocharged flat-six took a leap forward, delivering 462 hp.

The bucket seats were upholstered in leather, and standard equipment included front and side airbags, as well as power windows. Only the radio and air conditioning were installed on request. The range also included a Clubsport version with a bolted roll cage.

In autumn 2003, engineers revisited the GT2. While the Carrera GT supercar had just set new benchmarks, the queen of 911s received new electronic Motronic management mapping, as well as reworked pistons and cylinder liners.

Result: a gain of 21 hp and 20 Nm (15 lb-ft) of torque. These evolutions resulted in a top speed of 319 km/h (198 mph) and dazzling acceleration, with only 12.5 seconds required to reach 200 km/h (124 mph), compared to 12.9 seconds previously. A carbon rear spoiler made its appearance on the Clubsport version.

ENGINE: Flat-six biturbo
DISPLACEMENT: 3.6L / 220 in³
POWER: 462 hp at 5,700 rpm / 483 hp
TORQUE: 457 lb-ft from 3,500 to 4,500 rpm
GEARBOX: Six-speed manual
DIMENSIONS (L x W x H): 14' 7.2" x 6' 0.0" x 4' 2.2"
WEIGHT: 3,175 lbs / 3,131 lbs
TOP SPEED: 196 mph / 198 mph

Following measured evolutions, Porsche revisited its 911 GT2 in autumn 2003. Engineers succeeded in finding an additional 21 hp by modifying the electronic management mapping.

911 40 JAHRE
2003

Porsche's fondness for anniversaries was once again evident in 2003. On the 911's fortieth anniversary, the German manufacturer unveiled a special series at the Frankfurt Motor Show in September, simply named "40 Jahre," produced in a limited and numbered edition of 1963 examples. The entire series benefited from an identical and distinctive presentation. Featuring a metallic silver-gray paint developed for the famous Carrera GT, this anniversary version was recognizable at first glance by its sculpted front bumper from the Turbo, its distinctive side skirts, and its polished 18-inch wheels. A sunroof and bi-xenon headlights with dynamic range control completed the exterior design. Personalization continued in the interior with aluminum-style instrument surrounds, sport seats upholstered in dark gray leather, and trim in the same color as the bodywork. A combined radio CD player and a numbered "40 Jahre" plaque accompanied the bespoke finish.

Under the rear hood, the latest evolution of the 996's 3.6-liter flat-six was found, but the engineers had reworked it. The optional X51 kit enabled it to gain 25 horsepower, increasing power to 345 horsepower. To take full advantage of the increased performance – reaching 100 km/h (62 mph) in 5.4 seconds and a top speed of 290 km/h (180 mph) – the chassis was lowered by 10 mm. It benefitted from the PSM stability control system, also used in the Carrera 4, as well as a mechanical rear differential lock that improved rear wheel traction.

ENGINE: Flat-six
DISPLACEMENT: 3.6L / 219 in^3
POWER: 345 hp at 6,900 rpm
TORQUE: 274 lb-ft at 4,250 rpm
GEARBOX: Six-speed manual
DIMENSIONS (L x W x H): 14' 6.4" x 5' 9.7" x 4' 3.4"
WEIGHT: 3,131 lbs
TOP SPEED: 179 mph

ABOVE: Forty years of uninterrupted market presence was celebrated with a special series simply featuring the 911 logo on the engine cover, underlined by the "40 Jahre" signature.

RIGHT: The 3.6-liter flat-six engine of the 911 "40 Jahre" also received special treatment, featuring an optional kit that increased power to 345 hp.

RIGHT: The commemorative 911 fortieth anniversary version was recognizable by its metallic silver paint, borrowed from the Carrera GT supercar, and its sculpted front bumper from the 911 Turbo.

OPPOSITE: Produced in only 200 units, the 911 GT3 RS was primarily aimed at collectors and enthusiasts of track days.

911 GT3 RS
2003

During the second half of 2003, Porsche continued to exploit the advantage of special series by designing an RS (RennSport) version of the GT3. In a way, a super GT3. The objective assigned to the engineers was to provide a homologation base for the competition model and to enhance the sports car's performance, primarily benefiting efficiency on track. In practice, the GT3 RS's 3.6-liter block (381 hp) remained identical to that of the GT3 and the 911 GT1, which was victorious at the 1998 24 Hours of Le Mans. However, the fresh air intake was optimized to reduce inertia. A single-mass flywheel from the GT3 Clubsport completed the equipment. The transmission was revised to adapt to the constraints of competition. The gearbox gears were mounted on the shaft, rather than being machined into it, to allow for changing gear ratios according to the track configuration.

The chassis underwent new developments to lower the center of gravity. The anchor points of the steering linkages, the front suspension geometries, and the arms of the multi-link rear suspension were modified accordingly. Michelin Pilot Sport tires gave way to Pirelli P Zero Corsa System slicks re-cut for the road.

ENGINE: Flat-six
DISPLACEMENT: 3.6L / 220 in³
POWER: 381 hp at 7,400 rpm
TORQUE: 284 lb-ft at 5,000 rpm
GEARBOX: Six-speed manual
DIMENSIONS (L x W x H): 14' 6.4" x 5' 9.5" x 4' 2.0"
WEIGHT: 2,998 lbs
TOP SPEED: 190 mph

Approximately twenty kilograms (44 lbs) were saved through the use of carbon fiber for the front and rear hoods, mirror housings, a rear spoiler, and Recaro carbon-kevlar bucket seats. The radio and air conditioning were not standard equipment, but they were available as complimentary options. The suede steering wheel rim and roll cage completed the portrait of a truly extreme GT3. Porsche wanted to take the GT3 RS concept even further. More provocatively, the Porsche 911 GT3 RS was limited to only 200 examples, which helped maintain a high value and feed the myth of the Porsche 911. It was recognizable by its exterior decoration, inspired by that of the 911 2.7 Carrera RS.

BELOW: Inspired by the Clubsport spirit, the 996 GT3 RS dispensed with radio and air conditioning but featured a suede steering wheel rim and a roll cage.

GT3RS

Carrera
S GO 1100

VI

TYPE 997

—

2004–2012

Eliminating the flaws of the previous model and improving performance in terms of both road behavior and reducing polluting emissions constituted the priorities of the 997 program. With the sixth generation, the 911 had never worn such a diverse range of faces. The 997 family, which was available in 26 variants throughout its career, gained direct fuel injection and a dual-clutch gearbox with the mid-2008 facelift.

At the time of passing the torch to the sixth generation, the 996 had certainly broken all production records with 160,000 units sold, but its subdued forms never truly convinced. Bearing the internal designation 997, the new model was presented in May 2004. It brought back the oval headlights and wide hips that made 993 admirers fantasize. Three mm (.12 in) shorter but 48 mm (1.89 in) wider than its predecessor, the new 911 was offered from launch in Carrera and Carrera S versions, equipped respectively with a 3.6-liter and a new 3.8-liter engine. The models shared the same bodywork for the first time since 1977. The increase in track width went hand in hand with the mounting of flush wheels, whose diameter now reached 18 inches for the base version and 19 inches for the Carrera S, as well as variable-ratio steering.

The naturally aspirated flat-six of the Carrera produced 325 hp at 6,800 rpm, while that of the Carrera S reached 355 hp at 6,600 rpm. With a torque of 400 Nm (295 lb-ft), the Carrera S had 30 Nm (22 lb-ft) more than its smaller sister. These two engines were optionally paired with the new six-speed manual gearbox or the Tiptronic S five-speed automatic transmission.

The generational change also resulted in the adoption of a new interior giving more importance to instrumentation and the central infotainment screen. The remarkable driving position was achieved by moving the pedal cluster back 10 mm (.39 in), lowering the seats, and adjusting the steering wheel for both depth and height by almost 40 mm (1.57 in). Despite increased standard equipment, the weight did not increase, displaying 1,395 kg (3,075 lbs) and 1,420 kg (3,130 lbs), respectively. This was the result of weight reduction efforts on the suspensions, wheels, and engine cover, which was now made of aluminum. The ceramic brake option saved an additional 14 kg (31 lbs).

Unlike the Carrera, the Carrera S was equipped with PASM (Porsche Active Suspension Management), a controlled suspension with ten damping settings. It also featured a sport chassis with stiffer and 20 mm (.79 in) lowered suspensions, a limited-slip differential with differentiated locking, and the Sport Chrono Plus Package, which integrated a stopwatch that memorized lap times. The 20-second gain recorded by Walter Röhrl on the Nürburgring Nordschleife validated the progress made.

A few months later, the Cabriolet and the Carrera 4 version, recognizable by its 44 mm (1.73 in) wider rear fenders, returned to the catalog. In early 2007, the family expanded with a new Targa model. The glass roof still slid back, but the model was, for the first time in history, available solely with all-wheel drive with the 3.6 and 3.8-liter engines. Its structure, derived from that of the Carrera 4, used windshield pillars borrowed from the Cabriolet. These were extended by reinforcing bars housed in the aluminum roof arches. In 2008, as it passed its sixtieth birthday, Porsche revised the 911. It benefited from a new evolution of the flat-six, clearly emphasizing reduced fuel consumption and lower CO2 emissions.

RIGHT: Three years after its launch, the 911 Turbo adopted the 3.8-liter flat-six and direct fuel injection.

PREVIOUS TWO PAGES: With the sixth generation, the 911 reached a new level of integration with electronic aids.

While the Carrera's fuel consumption dropped by 0.7 liter/100 km, power increased by 20 hp (345 hp). The same was true for the Carrera S, whose engine consumed 0.4 liter/100 km less and gained 30 hp, reaching 385 hp, crossing the symbolic 100 hp/liter mark in the process. The primary source of this energy saving was the adoption of direct fuel injection. This technology fed a new, more compact six-cylinder engine, distinguished by an even shorter piston stroke. However, the main attraction of this mid-career facelift was the replacement of the old Tiptronic S five-speed automatic gearbox by a seven-speed robotic manual gearbox called PDK (Porsche-Doppelkupplungsgetriebe, or "dual-clutch gearbox"). Porsche had used this technology in competition in the 956 as early as 1983. With a largely overdriven seventh gear, the PDK gearbox also contributed to reducing fuel consumption. These evolutions were accompanied by LED daytime running lights, redesigned front and rear bumpers, and a center console now equipped with a touch screen.

Always launched with a slight delay compared to the Carrera, the 911 Turbo debuted in May 2006, featuring its main innovation: a 3.6-liter flat-six engine equipped with two variable geometry turbos. This allowed it to claim 480 hp and 620 Nm (457 lb-ft) of torque, from 5,000 to 6,000 rpm. With the Sport Chrono Package, an over boost increased this value to 680 Nm (502 lb-ft) for 10 seconds. Performance was exceptional: speed was limited to 310 km/h (192 mph) to preserve its tires at full load, but it accelerated from 0 to 100 km/h (62 mph) in just 3.9 seconds, the same time as the Carrera GT. The chassis adapted accordingly. The braking system was borrowed from the Carrera GT supercar, featuring active PTM (Porsche Traction Management) all-wheel drive that distributed torque to the four wheels based on grip, as well as an electromagnetically controlled clutch. A limited-slip differential could complete the optional equipment. Three years after its launch, the 911 Turbo benefited from a new engine derived from the A1 block that already powered the Carrera. With a displacement of 3.8 liters, this lighter and more compact flat-six also received direct fuel injection. Power now reached 500 hp and maximum torque of 650 Nm (479 lb-ft). Acceleration improved significantly: 3.4 seconds were enough to break the 100 km/h (62 mph) barrier. In turn, the Turbo was equipped with a PDK seven-speed robotic gearbox. The chassis gained a new rear differential, called "torque vectoring".

While all eyes were already turned towards the next generation, Porsche surprised 911 enthusiasts by unveiling a GTS version at the Los Angeles Auto Show in November 2010. This new model, which brought the number of 911 variants to twenty-six, resulted from the marriage of a widened Carrera 4 body and a 10 mm (.39 in)-lowered chassis with a flat-six engine boosted to 408 hp. Strictly rear-wheel drive, the GTS was lightened by approximately 50 kg (110 lbs), thanks to the removal of the rear seats, among other modifications.

OPPOSITE: Judged disappointing by GT3 enthusiasts, the limited weight reduction of 15 kg (33 lbs) was largely compensated by the presence of the 415 hp 3.6-liter flat-six, whose base came from the famous 911 GT1 engine.

911 GT3
2006

The new GT3 officially debuted at the Geneva Motor Show in March 2006. The formula that Porsche mastered perfectly was adapted to the 997 generation. The structure was that of the Carrera 4, but engineers fitted it with the narrow body of the Carrera 2. Aerodynamic efficiency was achieved through a faired underbody, a redesigned and adjustable rear spoiler, and a slit in the front hood to ensure air evacuation from the radiator. A hunt for superfluous kilos limited the weight gain to 15 kg (33 lbs) or 1,395 kg (3,075 lbs) compared to the 996 GT3. Customers appeared to be no longer willing to compromise on a minimum level of comfort, as the new model retained the radio and multimedia system with navigation.

The surprise was under the rear hood. The GT3 did not succumb to the Carrera S's 3.8-liter flat-six but retained the 3.6-liter, with power increased to 415 hp at 7,600 rpm. This progress was based on optimized air intake, an enlarged throttle body, and a new aluminum variable intake manifold. The flat-six was mated to a six-speed manual gearbox with ultra-short throws and close ratios. While some judged the performance insufficient – 100 km/h (62 mph) was reached in 4.3 seconds – road handling evolved profoundly thanks to the first-time use of an active PASM suspension with a sport mode. Not content with adopting the 4-liter RS's limited-slip differential, the GT3 benefited from a 30 mm-lowered chassis and the Carrera GT's switchable traction control.

At the 2009 Geneva Motor Show, the GT3 returned restyled. The subtle aesthetic evolution was based on a redesigned front bumper, headlight clusters, and air intakes, as well as an oversized rear spoiler that provided double the downforce compared to the previous version. The flat-six's displacement was increased to 3.8 liters, but the engine had nothing to do with the Carrera S. Devoid of direct injection but equipped with dry sump lubrication and the Variocam system, the engine produced 435 hp at 7,600 rpm.

With the 997 GT3, the structure was that of the Carrera 4, but engineers fitted it with the narrow body of the Carrera 2. The flat-six, combined with a six-speed manual gearbox with ultra-short throws and close ratios, allowed the 997 GT3 to break the 100 km/h (62 mph) barrier in 4.3 seconds.

ENGINE: Flat-six
DISPLACEMENT: 3.6L / 220 in³ / 3.8L / 232 in³
POWER: 415 hp at 7,600 rpm / 435 hp
TORQUE: 298 lb-ft at 5,500 rpm / 317 lb-ft at 6,250 rpm
GEARBOX: Six-speed manual
DIMENSIONS (L x W x H): 14' 7.0" x 5' 11.2" x 4' 2.4" / 14' 7.6" x 5' 11.2" x 4' 2.4"
WEIGHT: 3,075 lbs
TOP SPEED: 193 mph / 194 mph

GT3
AB-098-VH

911 GT3 CUP
2006

There was no question of changing a well-oiled formula. The 911 – intended to compete in the Supercup series and in the various national Carrera Cup championships – benefited from both the experience gained with the previous generation model and the studies of the 997 GT3. It was unveiled in March 2006, at the Geneva Motor Show.

Displacement remained unchanged at 3.6 liters, but the engineers still managed to gain 10 hp to reach 400 hp. It seems that the performance did not satisfy the participants of the various events, for in the 2008 model year, the 911 Cup returned with an increased power of 20 hp.

At the Frankfurt Motor Show in September 2009, Porsche presented a third version of the 997 Cup, derived for the first time from the GT3 RS. Entered into service from the 2010 season, this new model was recognizable by its body widened by 44 mm (1.73 in), its daytime running lights from the production 911, and its enormous rear spoiler from the GT3 Cup S competing in FIA GT3 events. Not content with being positioned higher, it exceeded its predecessor by 24 mm (.95 in) in width at 1.70 m (66.93 in), thus generating 40% additional downforce. The front bumper corresponded to that of the GT3 RS, but the aerodynamic lip extended 15 mm (.59 in) lower, to the benefit of increased front downforce. The characteristic air extractors and LED rear lights on the rear bumper came from the road version. The 2010 GT3 Cup was fitted with wider wheels, equipped with three-piece light alloy rims. PCCB carbon-ceramic brakes, reserved for the Supercup variant, reduced weight by 20 kg (44 lbs) compared to the composite cast iron discs of the Carrera Cup. Unibal joints appeared at the transverse arms, while the front and rear anti-roll bars were now adjustable in seven positions. Driving comfort was enhanced by an airier interior with an information screen in the steering wheel, now adjustable in height.

The main evolution was hidden under the rear hood. Engineers installed the 3.8-liter flat-six engine which produced 450 hp at 7,500 rpm while the maximum revs were set at 8,500 rpm.

Presented in early 2006 at the Geneva Motor Show, the Carrera Cup, designed based on the 997, entered service a few weeks later.

ENGINE: Flat-six
DISPLACEMENT: 3.6L / 220 in³ / 3.8L / 232 in³
POWER: 400 hp at 7,400 rpm / 420 hp / 450 hp at 7,500 rpm
TORQUE: N/A
GEARBOX: Six-speed sequential
WEIGHT: 2,601 lbs / 2,646 lbs
TOP SPEED: 190 mph

RIGHT: From the Carrera Cup, Porsche developed a Cup S version intended for FIA GT events.

BELOW: At Porsche, engineers perfectly mastered their subject: the interior of the 997 Carrera Cup prioritized ergonomics and driver safety.

911 GT3 RSR 2007

Approaching the end of its life cycle, the 996 RSR competition version did not disappoint, but it quickly showed its limits against increasingly strong competition. Thus, Porsche's competition department set to work, from the beginning of the 997 GT3 and Cup development program, on designing an RSR variant allowing the company's customers to defend their chances in the various GT championships and major endurance events on the calendar. Thoroughly reworked compared to the 996, the 997 represented an excellent development base. The engineers' work extended to all areas of the car, but aerodynamics, the engine, and weight distribution were given particular attention.

Entered in the GT2 category, the 997 GT3 RSR immediately distinguished itself from the model it replaced with a new front apron with hypertrophied air ducts – designed to improve braking system cooling. The presence of thin blades on the sides of the bumper testified to many hours spent in the wind tunnel to reduce drag and increase downforce. These devices worked in conjunction with air intakes in the front hood, a faired underbody, and an enormous rear spoiler overhanging another appendage reminiscent of the famous ducktail of the 1972 Carrera RS. The widened wheel arches and side skirts reinforced the GT3 RSR's arsenal.

Inside, the ambiance was typically racing with a bare chassis and a dashboard reduced to its simplest expression. The welded roll cage with cross-braces at the doors contributed to the car's rigidity. Over the seasons and changes in regulations, the GT3 RSR never stopped evolving. The flat-six engine underwent several modifications, increasing from 3.8 liters to 4 liters from the 2010 season. The gearbox benefited from experience gained with the RS Spyder. The RSR's performance allowed it to become the most successful GT in the world, winning numerous 24-hour classics as well as the GT2 category of the 2010 24 Hours of Le Mans.

Memorizing the operation of all controls onboard required some time for the drivers.

ENGINE: Flat-six
DISPLACEMENT: 3.8L / 232 in³ / 4.0L / 244 in³
POWER: 460 hp
TORQUE: 317 lb-ft at 7,250 rpm
GEARBOX: Six-speed sequential
WEIGHT: 2,690 lbs
TOP SPEED: 190 mph

BELOW: Impressive, the 997 GT3 RSR was distinguished by considerably widened tracks and an enormous rear spoiler overhanging another appendage reminiscent of the famous ducktail of the 1972 2.7 Carrera RS.

ABOVE: With the 997 GT3 RSR, Porsche Motorsport engineers focused their efforts on aerodynamics, which featured a new front apron and fins.

911 GT3 RS
2007

In order to comply with the homologation rules of the main GT championships and the demands of track day enthusiasts, Porsche developed an RS (RennSport) version of the GT3. Presented at the Frankfurt Motor Show in September 2007, this model differed from its sister by its 44 mm (1.73 in) wider rear fenders, an additional opening in the black front spoiler lip, and complete wheel carriers and split transverse arms on the rear axle.

Engineers managed to save 20 kg (44 lbs) by installing an adjustable carbon fiber rear spoiler, a synthetic rear window, and lightweight bucket seats. With a weight of 1,375 kg (3,031 lbs) and unchanged power of 415 hp, the power-to-weight ratio decreased slightly. For intensive track use, the flat-six was equipped with dynamic suspension mounts.

The GT3 RS gained 0.1 seconds in the 0-100 km/h (62 mph) sprint. The catalog included a Clubsport variant with a roll cage, pre-equipment for the main battery kill switch, fire extinguisher, and six-point harness.

Following the evolution of the GT3, the RS variant returned even more refined for the 2010 model year. It did not go unnoticed with its 26 mm (1.02 in) wider front fenders, larger exhaust tips, the RS logo on the wheel center lock nut, and its imposing carbon fiber rear spoiler mounted on aluminum supports. Larger intake manifolds and a new air filter allowed the new 3.8-liter engine to deliver 450 hp at 7,900 rpm and 430 Nm (317 lb-ft) of torque at 6,750 rpm. With the six-speed close-ratio gearbox and a weight reduction of 5 kg (11 lbs), the GT3 RS phase 2 passed the 100 km/h (62 mph) mark in 4 seconds. Inside, a sporty ambiance was de rigueur with bucket seats upholstered in fabric or Alcantara, a strap-operated door mechanism, and a roll cage bolted to the chassis.

For the 2010 model year, Porsche marketed a revised and improved GT3 RS, adopting 26 mm (1.02 in) wider front fenders and a new 450 hp 3.8-liter flat-six.

ABOVE: Collectors rejoiced when Porsche announced that its 997 GT3 would be offered in a RennSport (RS) version. The bodywork became more impressive, and performance significantly improved.

RIGHT: Sporty ambiance aboard the 997 GT3 RS, serving as the base for a race car. Equipment included lightweight bucket seats, a strap-operated door mechanism, and a roll cage bolted to the body.

ENGINE: Flat-six

DISPLACEMENT: 3.6L / 220 in³ / 3.8L / 232 in³

POWER: 415 hp at 7,600 rpm / 450 hp at 7,900 rpm

TORQUE: 298 lb-ft at 5,500 rpm / 317 lb-ft at 6,750 rpm

GEARBOX: Six-speed manual

DIMENSIONS (L x W x H): 14' 7.6" x 6' 0.9" x 4' 2.4"

WEIGHT: 3,031 lbs / 3,020 lbs

TOP SPEED: 193 mph

911 GT2
2008

At Porsche, less is more, meaning two-wheel drive equals more sensations. Following that philosophy, the brand continued the 911 GT2, the ultimate high-performance version of the 911 Turbo. Although the two models share many components, the GT2 stands apart with its front bumper featuring a pronounced aerodynamic splitter, 19-inch ten-spoke alloy wheels, rear bumper air vents, and a twin-deck rear wing incorporating lateral air scoops. These aerodynamic refinements reduce the drag coefficient to 0.32 (compared with 0.34 for the Turbo). At the same time, weight was brought down to 1,440 kg, without excessive compromise in equipment. As the marque's flagship, the GT2 still offers a degree of refinement: a high-quality interior trimmed in leather and Alcantara, and ultra-light bucket seats of new design using a carbon-fiber-reinforced composite structure. These folding seats allow access to the rear, which is stripped of seats and lined with carpeting.

The GT2 claims the title of the fastest and most powerful road-legal 911 ever built. Its 3.6-liter flat-six, derived from the 911 Turbo, is fitted with new variable-geometry turbochargers and runs higher boost pressure. Output reaches 530 hp at 6,500 rpm – 50 hp more than the standard Turbo – while torque peaks at 680 Nm between 2,200 and 4,500 rpm. Offered exclusively with a six-speed manual gearbox, the GT2 delivers exceptional performance: 0-100 km/h in 3.7 seconds and a top speed of 329 km/h. A redesigned intake system is paired with a lighter exhaust – the first road-legal Porsche to feature a titanium rear silencer and tailpipes.

A flagship of the brand, the 997 GT2 had the title of the fastest and most powerful 911 ever homologated for the road upon its appearance in 2008.

To heighten the senses, the GT2 incorporates a "launch control" system that enables explosive starts reminiscent of Formula 1. It automatically sets the engine to its optimal revs; the driver merely releases the clutch for a perfect getaway. Porsche Stability Management (PSM) keeps the GT2 composed by managing torque delivery to the road.

The chassis, lowered by 25 mm compared with the 911 Carrera, is derived from the 911 GT3. The PASM active suspension is stiffened, while braking is handled by standard carbon-ceramic discs with monobloc calipers, cutting unsprung weight by around 20 kg compared to cast-iron rotors. For track-day enthusiasts, PSM can be partially deactivated, and tire-pressure sensors monitor inflation continuously.

In autumn 2010, the GT2 returned in an RS version limited to 500 units. Its mission: to restore Porsche honor after Nissan's GT-R had claimed a record lap on the Nürburgring Nordschleife. With a 7 minute 18 second lap, Porsche set the record straight. To achieve this, the GT2 RS gained 90 hp for a total of 620 hp and underwent a strict carbon-fiber weight-reduction program that shaved 70 kg. The results were stunning: 0-100 km/h in 3.5 seconds, 0-200 km/h in 9.8, and a top speed of 330 km/h.

LEFT: Top-tier sports performance did not exclude a certain interior refinement. The high-quality finish was accompanied by comprehensive equipment.

ENGINE: Flat-six biturbo
DISPLACEMENT: 3.6L / 220 in³
POWER: 530 hp at 6,500 rpm / 620 hp at 6,500 rpm
TORQUE: 502 lb-ft from 2,200 to 4,500 rpm / 516 lb-ft from 2,250 to 5,500 rpm
GEARBOX: Six-speed manual
DIMENSIONS (L x W x H): 14' 7.9" x 6' 0.9" x 4' 2.6"
WEIGHT: 3,175 lbs / 3,020 lbs
TOP SPEED: 204 mph / 205 mph

ABOVE: Only two-wheel drive, a weight reduced to 1,440 kg (3,175 lbs), a demonic power of 530 hp, a torque of 680 (502 lb-ft), and a top speed of 330 km/h (205 mph): the 997 GT2 set new benchmarks.

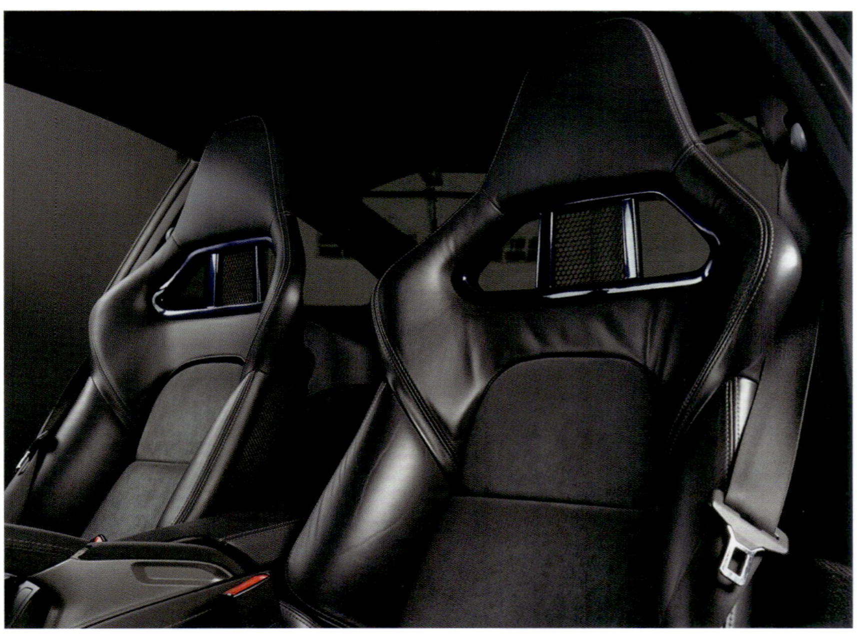

LEFT: The two ultralight bucket seats featured a new design, mixing plastic and carbon fiber in their structure.

911 SPORT CLASSIC 2009

Periodically, Porsche revisited its history. This was the case with the 911 Sport Classic presented at the Frankfurt Motor Show in 2009 and produced in a limited and numbered series of 250 examples. A product of the Porsche Exclusive department, this unique model largely referenced a glorious past, with its fixed ducktail rear spoiler, which evoked the famous 2.7 Carrera RS of 1972, and its mixed leather and woven textile upholstery, reminiscent of 911s from the 1960s.

Based on a Carrera S with rear tracks widened by 44 mm (1.73 in), the 911 Sport Classic was further distinguished by its mouse gray color (Sport Classic Gray), the double-bubble roof, the 20 mm (.79 in) lowered sport chassis, and 19-inch Fuchs wheels with five black spokes.

The 911 Sport Classic introduced the new 3.8-liter flat-six that would be found a few months later under the hood of the Carrera GTS. With a power output of 408 hp, this direct fuel-injection engine was paired with a six-speed manual gearbox.

The unique character of this limited series was reinforced by the following equipment: carbon-ceramic brake discs, limited-slip rear differential, PASM controlled suspension, Bose sound system, refined Nappa brown leather interior, and backlit "Sport Classic" door sills.

Presented at the 2009 Frankfurt Motor Show, the Sport Classic maintained the fascination for the 911 by reviving memories of the legendary 1972 2.7 Carrera RS.

ABOVE: Not content with bringing back the ducktail of the 911 2.7 Carrera RS, the Sport Classic introduced the 3.8-liter naturally aspirated flat-six engine destined for the Carrera GTS.

RIGHT: All identical, the 250 examples of the 2009 Sport Classic featured a refined interior in brown Nappa leather and houndstooth fabric.

ENGINE: Flat-six

DISPLACEMENT: 3.8L / 232 in^3

POWER: 408 hp at 7,300 rpm

TORQUE: 310 lb-ft between 4,200 to 5,600 rpm

GEARBOX: Six-speed manual

DIMENSIONS (L x W x H): 14' 6.8" x 6' 0.9" x 4' 2.8"

WEIGHT: 3,142 lbs

TOP SPEED: 188 mph

911 SPEEDSTER
2010

The economic and financial crisis, stemming from the bankruptcy of Lehman Brothers Bank in September 2008, led to a change in strategy in Zuffenhausen, against a backdrop of latent war between the Porsche and Piëch cousins, and a merger between Porsche and Volkswagen. Forecasting a drop in sales, product managers sought to maintain the brand's and various ranges' appeal. Thus, the idea of adding a Speedster to the 911 range was born. Presented at the Paris Motor Show in 2010, this fourth Speedster in the brand's history turned its back on the stripped-down nature associated with this architecture. Its production was limited to 356 units, in homage to the brand's first model. Priced almost twice as much as a 911 Carrera S Cabriolet, it shone neither in its lightness nor in its elegance. The bodywork suffered from a rather unaesthetic double rear hump under which the manually operated soft top was stored.

Under the guise of celebrating the twenty-fifth anniversary of the Porsche Exclusive personalization department, the 2010 Speedster featured a specific finish. It inherited a body widened by 44 mm (1.73 in) compared to the Carrera model, the 408 hp 3.8-liter flat-six from the Carrera GTS, the PDK dual-clutch seven-speed gearbox, PCCB (Porsche Carbon Ceramic Brakes) carbon-ceramic brakes, PASM variable damping shock absorbers, and Fuchs wheels inspired by those of the 911 Type G. Two colors were available: royal blue or white.

Developed in haste and contributing to supporting commercial activity during an economic crisis, the 997 Speedster was not renowned for its elegance.

ENGINE: Flat-six
DISPLACEMENT: 3.8L / 232 in^3
POWER: 408 hp at 7,300 rpm
TORQUE: 310 lb-ft from 4,200 and 5,600 rpm
GEARBOX: PDK seven-speed
DIMENSIONS (L x W x H): 14' 10.3" x 6' 0.9" x 4' 2.6"
WEIGHT: 3,395 lbs
TOP SPEED: 190 mph

ABOVE: A nod to Porsche's first car, the 997 Speedster was produced in only 356 examples. Its 408 hp 3.8-liter flat-six came from the Carrera GTS.

RIGHT: Standard equipment on the 997 Speedster included a manually operated soft top, a seven-speed PDK dual-clutch gearbox, and PCCB carbon-ceramic brakes.

911 GT3 R HYBRID
2010

Faced with ecological pressure, which primarily thrived in Europe and caused public authorities to tighten regulations, Porsche experimented with a multitude of technologies to reduce polluting emissions. In this regard, competition represented a valuable laboratory.

At the Geneva Motor Show in March 2010, the manufacturer unveiled a study of a hybrid 911 competition car. The 4-liter 480 hp flat-six was joined by two electric motors installed at the front, each delivering power of 60 kW (82 hp). They were powered by a flywheel accumulator located in the passenger compartment, near the driver. This placement contributed to maintaining a balanced mass distribution. The accumulator recharged during braking phases by recovering kinetic energy. The driver could then request the energy stored in the accumulator during acceleration out of corners or during overtaking maneuvers. The flywheel could provide up to 120 kW (164 hp) of kinetic energy to drive the two electric motors. This additional power was available for approximately 6 to 8 seconds after each charging phase.

The technology developed by Porsche was distinguished by greater fuel efficiency. Lower consumption reduced the fuel tank capacity and the number of pit stops.

After numerous tests, the GT3 R Hybrid was well on its way to winning the 24 Hours of Nürburgring in May 2010, one of the most significant endurance events reserved for GT cars. But 1 hour and 45 minutes before the checkered flag, the electrified GT retired due to a broken valve spring. At the end of the 2010 season, the experimental 911 was seen again on the starting grid of the ILMC (Intercontinental Le Mans Cup) round of the Petit Le Mans and then the Zhuhai 1000 km event. In China, it finished sixth and first in GT.

ENGINE: Flat-six
DISPLACEMENT: 4.0L / 243.9 in^3
POWER: 473 hp at 7,250 rpm + two electric motors of 81 hp (combined 635 hp)
TORQUE: N/A
GEARBOX: Six-speed sequential
DIMENSIONS (L x W x H): 14' 10.3" x 6' 4.9" x 4' 1.6"
WEIGHT: 2,866 lbs
TOP SPEED: 180 mph

A surprise at the Geneva Motor Show in March 2010, the 911 GT3 R Hybrid revolutionized the world of competition with its 4-liter 480 hp flat-six combined with two 82 hp electric motors.

The following year, the GT3 R Hybrid again took the start of the Eifel classic with improved technology. The power of the electric motors was increased to 75 kW, allowing a boost of 200 hp, while a restrictor reduced the power of the flat-six to 448 hp. Undoubtedly one of the fastest cars on the grid, the GT3 R Hybrid was delayed by differential problems that prevented Jörg Bergmeister, Richard Lietz, Marco Holzer, and Patrick Long from fighting for victory.

Engineers focused on installing the various components of the hybrid system for optimal mass distribution.

The interior, similar to that of the Cup and RSR competition versions, was nevertheless distinguished by the installation of the flywheel accumulator on the passenger side.

At the forefront of innovation, Porsche developed a 911 with a hybrid engine in 2010, which was entered into several endurance events.

OPPOSITE: A special series intended for buyers of the 918 Spyder plug-in hybrid supercar, the 911 Turbo of the same name featured illuminated signatures on its side skirts. At Porsche, attention to detail was illustrated by the instrument needles, which matched the fluorescent yellow color of the brake calipers.

911 TURBO S EDITION 918 SPYDER 2011

What was the best way to await the delivery of a 918 Spyder? Porsche decided in 2011 to offer every customer who had ordered the famous plug-in hybrid supercar the opportunity to acquire a 911 Turbo S with an "Edition 918 Spyder" finish. No fewer than 149 enthusiasts were seduced by this special model, available as a coupe or cabriolet.

This special version inherited many features from the 918 Spyder, including the distinctive silver paint and numerous fluorescent yellow elements such as the carbon-ceramic brake calipers, the carbon mirror trim, the embroidered monograms on the adaptive sport seats and the stitching on the leather upholstery, the instrument cluster needles, the multimedia system screen, and the illuminated door sills. The plaque indicating the limited edition bore the order number, a critical detail.

Under the hood, there was the 3.8 twin-turbo flat-six with 530 hp and 700 Nm (516 lb-ft) of torque. It was mated to a seven-speed PDK dual-clutch gearbox operated by paddle shifters on the steering wheel. The standard equipment of the 911 Turbo also included cruise control and the Sport Chrono Package.

ENGINE: Flat-six biturbo
DISPLACEMENT: 3.8L / 231.9 in^3
POWER: 523 hp at 6,750 rpm
TORQUE: 516 lb-ft at 4,250 rpm
GEARBOX: PDK seven-speed
DIMENSIONS (L x W x H): 14' 7.2" x 6' 0.9" x 4' 3.2"
WEIGHT: 3,638 lbs
TOP SPEED: 196 mph

No fewer than 149 enthusiasts who had ordered a 918 Spyder were seduced by this limited-edition model, available as a coupe or cabriolet.

Edition 918 Spyder

3
4
5
6
7
8
turbo S
80
120
P
R
N
D

OPPOSITE: A few weeks before revealing its 911 generation 991, Porsche made headlines by presenting, at the end of April 2011, the most powerful GT3 in its history.

911 GT3 RS 4.0
2011

The news caused a sensation. On April 28, 2011, a few weeks before revealing a new generation of the 911, Porsche surprised its followers by presenting a final version of the GT3 equipped with the most powerful naturally aspirated engine ever developed. Produced in a limited series of only 600 examples, all numbered and in white, this model, named GT3 RS 4.0, borrowed its features from the world of racing.

The adoption of a carbon fiber front hood and front fenders, composite rear windows, lightweight bucket seats, and door pull straps, along with the removal of the rear seats and sound insulation, resulted in a weight reduction to 1,360 kg (2,998 lbs). The power-to-weight ratio of 2.72 kg (6 lbs)/ch was achieved by installing a new flat-six engine whose displacement was increased to 4 liters. Derived from racing, this engine received forged pistons, titanium connecting rods, and a crankshaft identical to that of the 911 GT3 RSR. As a result, it delivered 500 hp at 8,250 rpm and 460 Nm (339 lb-ft) of torque at 5,750 rpm. Equipped solely with a six-speed manual gearbox, with its closer gear ratios the RS 4.0 reached 100 km/h (62 mph) in 3.9 seconds and crossed the 200 km/h (124 mph) mark in 11.9 seconds.

The work on the engine and weight was combined with extensively revised aerodynamics and a lowered and improved chassis. Featuring front side deflectors, bumper lips, a large rear wing, 19-inch wheels with central locking, carbon-ceramic brake discs, and a dual central exhaust outlet, the RS 4.0 was immediately recognizable. All the improvements made were validated by a lap time of 7 minutes 27 seconds on the Nürburgring Nordschleife.

ENGINE: Flat-six
DISPLACEMENT: 4.0L / 243.9 in³
POWER: 500 hp at 8,250 rpm
TORQUE: 339 lb-ft at 5,750 rpm
GEARBOX: Six-speed manual
DIMENSIONS (L x W x H): 14' 7.6" x 6' 0.9" x 4' 2.4"
WEIGHT: 2,998 lbs
TOP SPEED: 193 mph

LEFT: The last limited-series version of the 997 generation, the GT3 4.0, was available with either carbon racing bucket seats or electrically adjustable leather and Alcantara semi-bucket seats.

ABOVE: Previously reserved for racing models, the 4-liter flat-six was the highlight of this GT3, produced in only 600 units. It had a power output of 500 hp.

PORSCHE
S GO 5005

MONTANA

VII

TYPE 991

–

2011–2018

With increasing refinement, but in continuity with previous versions, the seventh-generation 911 underwent two successive revolutions. The first accompanied its launch when to the entirely new chassis promising transformed road behavior, the 991 added a more luxurious interior and engines that were both powerful and economical. The second, which arrived in the middle of its career, was marked by the adoption of a new supercharged, downsized engine. With 217,930 units, the 991 set a new production record.

Since the 996, each generation of the 911 had returned to a lifespan similar to most production vehicles. Internally named "991," the seventh generation of Porsche's sports car made its world premiere at the Frankfurt Motor Show in September 2011. It represented the most significant technical advance in the model's history. Under the direction of Michael Mauer, the design aimed to reproduce the timeless silhouette, but the platform was completely new. The 991 was more voluminous but also more balanced due to reduced overhangs. The wheelbase was lengthened by 100 mm (3.93 in) for an overall increased length of 56 mm (2.2 in). While the width was identical to the 997, the front had a widened track of 52 mm (2.05 in) or 46 mm (1.81 in) on the 350 hp Carrera. The appearance improved in tension and elegance, especially at the rear, which featured slim taillights. Made of aluminum and steel, the lightweight body, 80 kg (176 lbs) lighter and 20% more torsionally rigid, compensated for the extra weight due to new comfort and safety features. The result was an overall weight saving of 45 kg (99 lbs).

Inside, the change was even more spectacular with improved ergonomics around a center console borrowed from the Panamera sedan and a more luxurious presentation. Interior space increased by 25 mm (.98 in) at the front, 6 mm (.24 in) at the rear where space could accommodate passengers in a pinch or luggage after folding down the backs of the two small seats. The list of equipment illustrated the 911's entry into the 21st century: PCM infotainment system with real-time traffic information, electric power steering, standard Stop & Start, coasting mode, and refined thermal management. Engineers had managed to reduce fuel consumption – no version exceeded 10 liters per 100 km – while increasing performance. The displacement of the 911 Carrera was reduced to 3.4 liters, but power increased by 5 hp to 350 hp at 6,400 rpm. Meanwhile, the 3.8-liter Carrera S was boosted to 400 hp at 7,000 rpm. In addition to the PDK transmission, the 911 debuted a seven-speed manual gearbox. Road handling gained in docility and efficiency thanks to perfected chassis components, as well as active anti-roll (PDCC – Porsche Dynamic Chassis Control) and adaptive damping (PASM) combined with a host of electronic aids.

While it had found its way back to success with its entire range, Porsche did not abandon its DNA. The usual body variants and versions of the 911 were reintroduced: Cabriolet, Carrera 4, Turbo (coupe and cabriolet), and Targa. In its fortieth year, the 911 Turbo in turn succumbed to the 991 designation and its host of innovations. The rear fenders widened by 28 mm (1.1 in) compared to the Carrera 4. From its launch, the 3.8-liter twin-turbo with variable geometry was available in two versions: 520 hp (Turbo) and 560 hp (Turbo S). The manual gearbox disappeared in favor of the PDK seven-speed dual-clutch automated transmission. Performance took a new leap forward (0 to 100 km/h (62 mph) in 3.2 seconds for the Turbo), while fuel consumption dropped by 16%, especially with the automatic engine cut-off at a standstill.

The increase in performance justified the development of a new

RIGHT: With the seventh-generation 991, the interior became more luxurious, spacious, and ergonomic. The presentation borrowed from the Panamera sedan featured a new infotainment system.

PREVIOUS TWO PAGES: Constructed on a new platform, the seventh-generation evolved in every direction.

active rear spoiler with three positions. It worked in conjunction with the front spoiler, which deployed in three segments using pneumatic control. To the evolutions seen on the 991, the Turbo added the rear-wheel steering from the GT3, turning in the opposite direction by 2.8 degrees below 50 km/h (31 mph) and in parallel by 1.5 degrees above 80 km/h (50 mph), benefiting agility and stability during rapid trajectory changes. The test on the Nürburgring Nordschleife demonstrated that the Turbo S had dropped to a 7 minutes 24 seconds lap.

The 911 celebrated its fiftieth anniversary when the Targa model appeared. It brought back the metallic roll bar that equipped the first model of its kind in 1966. The very complex roof mechanism opened or closed in twenty seconds. The rear window and part of the engine cover lifted and tilted back as a single unit while the roof canvas emerged from its housing and positioned itself above the cockpit. Heavier by 40 kg (88 lbs) compared to the cabriolet, the Targa was only offered with all-wheel drive, associated for the first time with rear torque vectoring (PTV – Porsche Torque Vectoring).

At the Los Angeles Auto Show in 2014, the curtain lifted on a new GTS with a boosted 3.8-liter engine delivering 430 hp and available in four versions: coupe or cabriolet, two or four-wheel drive. The Targa joined the GTS family in January 2015. The Carrera 4 body combined with a 10 mm (.39 in) lowered chassis, active suspension, and 20-inch wheels with center lock. The exclusive interior was upholstered in Alcantara.

Causing astonishment at the Frankfurt Motor Show in 2015 during the mid-career facelift, the 911 Carrera abandoned its naturally aspirated flat-six for a turbocharged and downsized engine. This marked the end of an era for purists who flocked to dealerships to buy the last models. 911 enthusiasts did not lose out from the change. While displacement dropped to 3 liters, power increased by 20 hp, and torque by 60 Nm (44 lb-ft), for both versions. The chassis, lowered by 10 mm (.39 in), could now benefit from rear-wheel steering. Fuel consumption did not exceed 7.7 liters/100 km (Carrera S), corresponding to CO_2 emissions of 174 g/km. With the PDK gearbox, the Carrera S went below the 4-second mark for the first time (3.9) to reach 100 km/h (62 mph). This revolution was accompanied by a more refined character, a new touchscreen, and a four-point light signature reminiscent of the 919 Le Mans prototype. Gradually, the entire 911 family converted to the twin-turbo flat-six. Meanwhile, the 911 Turbo regained its distance from the Carrera with its 3.8-liter twin-turbo giving 20 hp for both versions.

In early 2017, as it prepared to cross the millionth unit produced, the 911 evolved further. Five new GTS variants were launched around the 3-liter twin-turbo flat-six boosted to 450 hp. The Targa 4S remained the only model in the lineup to be offered only with all-wheel drive.

911 CLUB COUPĒ
2012

At Porsche, every event or anniversary was an opportunity to design a special series. To commemorate the 60th anniversary of the first Porsche club, the German manufacturer designed a special edition car. Only 13 examples were produced, referring to the thirteen members who founded the club in 1952. The Zuffenhausen Museum kept the thirteenth, so Porsche only sold 12 units of this 911, simply named "Club Coupé."

The brand innovated by limiting purchase to members of official clubs worldwide. To select among the 181,000 members spread across the 634 existing clubs around the globe, Porsche organized a lottery that selected fifty applications. The first twelve were contacted by the relevant club federation at the factory's request to obtain final approval. Once the list was established, the twelve privileged individuals were received at the factory from December 9 to 11 (9-11), the twelfth month of the year.

The vehicles were all identical, except for the plaque bearing the name of each owner. The engineers started with a Carrera S to create this Club version featuring dark green paint – Wolfgang Porsche's favorite color – a polished aluminum window surround, 20-inch Sport Techno wheels in the same color as the body, and yellow calipers for the carbon-ceramic braking system. The Sport Design Package and the ducktail spoiler – a nod to the 1972 2.7 Carrera RS – completed the equipment of this model.

Borrowed from the 991 generation, the refined interior was distinguished by the brown leather upholstery, in harmony with the exterior color. The ignition key was in the same color as the vehicle and the illuminated door sills displayed the 911 Club Coupé logo accompanied by the Porsche clubs' sixtieth anniversary logo. Around a sport chassis lowered by 20 mm (.79 in) and a specific front bumper, this model benefited from an engine kit increasing the power of the 3.8-liter flat-six to 430 hp.

The thirteen Club Coupé versions shared the same refined interior: brown leather upholstery and steering wheel with white stitching, brown carpets, and an ignition key in the same color as the body.

ENGINE: Flat-six
DISPLACEMENT: 3.8L / 231.9 in^3
POWER: 424 hp at 7,500 rpm
TORQUE: 325 lb-ft at 5,750 rpm
GEARBOX: PDK seven-speed
DIMENSIONS (L x W x H): 14' 8.8" x 5' 11.2" x 4' 2.6"
WEIGHT: 3,120 lbs
TOP SPEED: 188 mph

ABOVE: In 2012, Porsche celebrated the sixtieth anniversary of its first club by launching a series of thirteen identical coupés based on the 991 Carrera S, equipped with the 435 hp engine kit.

RIGHT: Twelve vehicles were allocated by lot among Porsche club members worldwide, and the thirteenth example was intended for the brand's museum.

911 # GT3 2013

When he presented the third generation of the GT3 at the Geneva Motor Show in March 2013, Jürg Jünger, in charge of coordinating this project, did not hesitate to use the term "revolution" to define the concept behind the development of the sportiest 911. "As it celebrates its fourteenth birthday, the GT3 is going through its adolescence. Its voice became deeper, its behavior changed, but it remained fundamentally the same," Jünger ventured to emphasize the transformations made. Based on the 991, engineers had concocted a formidable machine, equaling the time of the Carrera GT supercar on the Nürburgring Nordschleife.

The most significant change concerned the abandonment of the manual gearbox in favor of a PDK seven-speed dual-clutch automatic transmission. This resulted in a weight gain of around 35 kg (77 lbs) compared to the 997 GT3, which the technicians sought to curb by reducing the weight of the bare body by 13% through extensive use of aluminum for the roof, fenders, hood, and doors. In the end, rigidity increased by 25% compared to the previous model. At the same time, many components were strengthened, not only to handle increased performance but also to account for the higher stresses they would have to withstand on the track. The rear fenders were widened by 44 mm (1.73 in) compared to a Carrera S, and the suspension was changed.

With the lowered chassis, modified axle geometry, stiffened settings, PASM adaptive damping, and nine and twelve-inch-wide wheels with a twenty-inch diameter, the GT3 laid the foundations for increased efficiency, to which was added, for the first time, rear-wheel steering with a steering angle of 1.5 degrees, benefiting stability and agility. The road links allowed the car to exploit the naturally aspirated flat-six, which was not the 4-liter from the Metzger lineage – named after the engine builder in charge of racing engines for several decades – but a derivative of the 3.8-liter from the Carrera S. With forged aluminum pistons, titanium connecting rods, and valve actuation by rocker arms, it was rated at 475 hp at 8,250 rpm. As it changed generation, the dimensions of the rear wing were increased, and the bumpers incorporated gaping air intakes.

With the 991 generation, the GT3 sports car abandoned the manual gearbox in favor of a PDK seven-speed dual-clutch automatic transmission, which was faster and more economical.

RIGHT: Presented in Geneva in March 2013, the third GT3 in Porsche's history marked a significant evolution of the sporty 911 concept, matching the Carrera GT supercar's time on the Nürburgring Nordschleife.

BELOW: Developed from a Carrera S, the GT3's road handling was significantly improved with the 44 mm (1.73 in) wider rear fenders, an oversized spoiler, and, for the first time, rear-wheel steering.

Although the second version of the GT3, which appeared in 2017, converted to the 4-liter flat-six, the first version was content with the 3.8-liter, boosted to 475 hp.

Four years later, the 911 GT3 returned in a revised version. It inherited the GT3 Cup's engine. Displacement was increased to 4 liters, and power increased by 25 hp to 500 hp due to a stiffer crankshaft mounted on wider bearings, plasma-coated cylinders, and valve lash maintained by factory-calibrated shims. The redline was set at 9,000 rpm.

In response to demands from enthusiasts, the six-speed manual gearbox was available again. It was coupled to a mechanical limited-slip rear differential, as opposed to an electronically controlled differential with the PDK gearbox combined with the "Torque Vectoring" function designed to facilitate the 911's entry into corners. Among the improvements over the 2013 GT3 was a spoiler positioned 20 mm (.79 in) higher and a re-tuned chassis for better body movements.

The two versions of the 911 GT3 Cup were developed from the GT3. The first, which appeared in 2013, delivered 460 hp and debuted paddle shifting. The second's displacement increased to 4 liters and its power climb to 485 hp. The two versions of the GT3 Cup were produced in 673 and 737 units respectively.

ENGINE: Flat-six
DISPLACEMENT: 3.8L / 4.0L (231.9 in³ / 243.9 in³)
POWER: 469 hp at 8,250 rpm / 493 hp
TORQUE: 325 lb-ft at 6,250 rpm / 339 lb-ft at 6,000 rpm
GEARBOX: PDK seven-speed
DIMENSIONS (L x W x H): 14' 11.0" x 6' 0.9" x 4' 2.0"
WEIGHT: 3,153 lbs / 3,115 lbs
TOP SPEED: 196 mph / 199 mph

S GO 5075

OPPOSITE: The result of extensive wind tunnel research, the aerodynamics of the 991-generation RSR incorporated a very large rear diffuser that works in concert with the enormous rear wing borrowed from the 919 Hybrid, the LMP1 racing prototype at Le Mans.

911 RSR
2013

In the fiftieth year of the 911, the Motorsport department unveiled a new model intended to succeed the 911 GT3 RSR in the World Endurance Championship. With 450 hp, the 4-liter flat-six was mated to a new six-speed automated gearbox controlled by paddle shifters on the steering wheel. Engineers focused their efforts on reducing and redistributing weight, as well as lowering the center of gravity. Carbon fiber was used to make the fenders, hoods, doors, wheel arches, rear wing, dashboard, and center console. Well-designed, the RSR achieved multiple victories during its career, including a double in the GTE-Pro category at the 2013 24 Hours of Le Mans.

The entry into the LMGTE (Le Mans Grand Touring Endurance) category in endurance races led Porsche to design a new racing 911 in 2016. Frank-Steffen Walliser, head of Porsche Motorsport, said that "it was the biggest evolution in the model's entire history." Engineers started from a blank slate to develop the new 911 RSR. Everything was new: the suspension, the body structure, the aerodynamic concept, the engine, and the transmission. They chose the naturally aspirated 4-liter flat-six engine with four valves per cylinder and direct injection, but it was positioned in front of the rear axle, in a mid-rear configuration. Depending on the size of the intake restrictor, it delivered nearly 510 hp. Power was transmitted to the rear wheels via a six-speed sequential gearbox with a magnesium casing. Gear changes were made by paddle shifters on the steering wheel.

Ready for the start of the 2017 season, this model, tasked with defending Porsche's colors in the GT category, was distinguished by a reduction in the number of carbon fiber body elements to reduce replacement times. The 911 RSR was the first Porsche GT race car to integrate the radar-assisted collision warning system: the "Collision Avoidance System."

ENGINE: Flat-six
DISPLACEMENT: 4.0L / 4.0L (243.9 in³ / 244.1 in³)
POWER: 454 hp at 7,800 rpm / 503 hp
TORQUE: 332 lb-ft at 7,250 rpm / n/a
GEARBOX: Six-speed sequential
DIMENSIONS (L x W x H): 14' 11.4" x 6' 8.4" x n/a
WEIGHT: 2,745 lbs / 2,740 lbs
TOP SPEED: N/A

ABOVE: The RSR was the first Porsche GT to integrate the radar-assisted collision warning system. The bucket seat was fixed, and the driving position was adjusted by regulating the pedal cluster.

LEFT: To commemorate the German brand's seventieth anniversary, the two 911 RSRs participating in the 2018 24 Hours of Le Mans featured liveries of models that had previously raced at Sarthe.

911 50
2013

If there was one anniversary that the Zuffenhausen manufacturer would not have missed for anything, it was the jubilee of its 911. A collector's version was produced, developed from the widened Carrera 4 chassis, but transmitting the 400 hp of the Carrera S's 3.8-liter flat-six only to the rear wheels. The fiftieth-anniversary edition was presented at the Frankfurt Motor Show in 2013 and was produced in only 1963 numbered units. Three colors were offered: Graphite Gray, Geyser Gray, and Black. This coupé brought back the famous Fuchs wheels, in a 20-inch variant with polished spokes. In addition, the special engine grille and the aluminum trim highlighting the side windows further differentiated this version from other models in the range. The interior also drew its references from the past. The seats featured a mix of leather and houndstooth fabric upholstery – reminiscent of the upholstery of the 1960s – but an all-leather interior remained an option.

The white and red "911 50" two-tone monogram appeared on the headrests, the aluminum door sills, and the tachometer. The Jubilee 911 received the benefit of the sport chassis associated with PASM active suspension management. Like a standard range model, it was sold with a seven-speed manual or PDK automatic gearbox.

ABOVE: For once, Porsche did not choose the green color for its special series celebrating the 911's fiftieth anniversary. However, the mixed leather and fabric upholstery revived the houndstooth pattern of the 1960s.

ENGINE: Flat-six
DISPLACEMENT: 3.8L / 231.9 in³
POWER: 395 hp at 7,400 rpm
TORQUE: 325 lb-ft at 5,600 rpm
GEARBOX: PDK seven-speed automatic / seven-speed manual
DIMENSIONS (L x W x H): 14' 8.8" x 6' 0.9" x 4' 3.2"
WEIGHT: 3,241 lbs
TOP SPEED: 186 mph

OPPOSITE (BOTH): Developed from the Carrera S version, the Jubilee 911 inherited the 3.8-liter flat-six with 400 hp. It was mated to either a seven-speed manual or a seven-speed PDK automatic transmission. In reference to its presentation year, the "50 years" special series was produced in only 1963 numbered units. It was recognizable by its special engine grille and the red logo on the rear bumper.

PORSCHE
S GO 9115
Porsche Museum

PORSCHE
911 50

911 CARRERA S MARTINI RACING 2014

The partnership between Porsche and the Martini spirits brand began in the late 1960s with the arrival of the 917 sports prototypes. This association was rewarded with numerous victories, both in class and endurance races, notably at the 24 Hours of Le Mans. On the eve of the 2014 Le Mans race, Porsche commemorated the partnership between the two companies with a special version of the 911 Carrera S, dubbed the "Martini Racing Edition."

Limited to 80 units and sold only in certain European countries as well as in China, Japan, and Latin America, this model was adorned with the Martini livery of the Porsche 935 Moby Dick that raced at Le Mans in 1978. For some markets, Porsche simply provided the decoration kit in the colors of the famous spirits brand. The 911 Carrera S Martini Racing was available, optionally, in white or black. It was distinguished from the production model by a front bumper with an aerodynamic lip and a fixed rear spoiler. Upon opening the doors, passengers were greeted by stainless steel door sills bearing the inscription "911 Carrera S – Martini Racing Edition" illuminated in red. The finish included a sport steering wheel, the Sport Chrono Package, the "Martini Racing Edition" logo on the dashboard, and electrically adjustable black leather sport seats.

Commemorating a long-standing sporting partnership rewarded with numerous endurance victories, the "Martini Racing Edition" version of the 911 features a distinctive finish in the brand's livery.

ABOVE: Developed from the 400 hp 911 Carrera S, the "Martini Racing Edition" version was produced in only 80 examples. It was available in two colors: white or black.

LEFT: For the "Martini Racing Edition" version, the designers contented themselves with taking the interior of a Carrera S equipped with the Sport Chrono Package and electric sport seats.

ENGINE: Flat-six

DISPLACEMENT: 3.8L / 231.9 in³

POWER: 400 hp at 8,250 rpm

TORQUE: 325 lb-ft at 5,600 rpm

GEARBOX: PDK seven-speed

DIMENSIONS (L x W x H): 14' 8.8" x 5' 11.2" x 4' 3.0"

WEIGHT: 3,241 lbs

TOP SPEED: 189 mph

911 GT3 R
2015

Unveiled in May 2015, the GT3 RennSport – the competition version of the GT3 – pushed the boundaries of the 911's architecture even further. Engineers succeeded in reducing weight, lowering the center of gravity and improving weight distribution.

There was no question of Porsche resting on its laurels. In May 2015, the Motorsport department unveiled its new weapon in the GT3 category: the GT3 R (for RennSport). Starting with the GT3 RS, engineers worked in several directions: reducing weight and improving mass distribution, lowering the center of gravity, enhancing aerodynamic efficiency, reducing fuel consumption, optimizing safety, and lowering maintenance costs. Like its production sibling, the GT3 R received the double-bubble roof, and the wheelbase was lengthened by 83 mm (3.27 in) compared to the previous generation.

On the scale, the new 911 customer race car weighed 1,220 kg (2,690 lbs) due to the use of carbon fiber reinforced plastic (CFRP) for the roof, front hood and fairing, wheel arches, doors, bumpers, and rear hood. Similarly, all windows, and even the windshield for the first time, were made of polycarbonate. The GT3 R featured an impressive aerodynamic package that had been tested in a wind tunnel. Air vents in the front fenders increased downforce on the front axle, while the rear wing – measuring two meters wide and forty centimeters deep – contributed to aerodynamic balance. The GT3 R borrowed the central radiator concept from its big sister, the 911 RSR. The removal of side radiators contributed to good mass distribution.

The naturally aspirated 4-liter flat-six engine was installed under the rear hood. With four valves per cylinder and direct fuel injection operating at pressures of 200 bar, this engine produced 500 hp. The driver changed gears using paddle shifters on the steering wheel. Safety was also a priority: the fuel tank was now equipped with a fuel cut-off valve. In 2018, the GT3 R underwent a new evolution, increasing power to 550 hp and further improving the aerodynamic concept.

In 2018, the Porsche Motorsport department presented a modified 911 GT3 R. Aerodynamics were optimized and the flat-six delivered 550 hp.

ENGINE: Flat-six
DISPLACEMENT: 4.0L / 244.1 in^3
POWER: 493 hp
TORQUE: N/A
GEARBOX: Six-speed sequential
DIMENSIONS (L x W x H): 15' 1.3" x 6' 6.8" x n/a
WEIGHT: 2,690 lbs
TOP SPEED: 186 mph

The edition of the 911 GT3 R, which appeared in 2018, paid attention to driver comfort. For the first time, the interior gained air conditioning.

911 GT3 RS
2015

The escalation of power and performance seemed to know no pause at Porsche. At the Geneva Motor Show in 2015, the 991 generation once again brought together its two sporting labels: GT3 and RS. This new cocktail was certainly flavorful. To earn the RS label, the GT3 used proven recipes: increased displacement, weight reduction, optimized aerodynamics, and running gear. Inherited from the special series GT3 RS 4.0 of 2011, the naturally aspirated flat-six was increased to 4 liters to deliver 500 hp at 8,250 rpm and a maximum torque of 460 Nm (339 lb-ft) at 6,250 rpm. This engine borrowed features from racing employing dry sump lubrication, titanium connecting rods, lightweight valve actuation by rocker arms, and a steel alloy crankshaft. At the same time, the mass was reduced by 10 kg (22 lbs) with the use of magnesium for the roof, carbon fiber reinforced plastic for the hoods and spoilers, and polycarbonate for the side and rear windows. The GT3 RS thus weighed in at 1,420 kg (3,131 lbs), which was 60 kg (132 lbs) more than its predecessor on the 997 chassis, but the new model was equipped with the PDK dual-clutch seven-speed transmission and the rear-wheel steering system debuted by the Turbo. The result: the GT3 RS boasted a flattering power-to-weight ratio of 2.8 kg (6.2 lbs)/hp. Decidedly very sophisticated, this RS benefited from a "clutch-off" function on the gearbox by pulling both paddle shifters on the steering wheel and a "Pit Speed" mode activating the speed limiter in the pits.

Based on the Turbo body, with its rear-fender air intakes, the 2015 GT3 RS drew extensively from the world of racing. Sitting on tires as generous as those of the 918 Spyder and mounted on center-lock wheels, the new super sports car in the GT3 family featured track widths widened by 50 mm (1.97 in) at the front and 72 mm (2.83 in) at the rear, compared to the Carrera model. With its aerodynamic lip on the front bumper, the louvered vents in the front fenders, and its three-position adjustable rear wing, downforce was doubled compared to the previous version. The 2015 GT3 RS's progress is illustrated by its 9-second advantage over the Carrera GT on the Nürburgring's large circuit. It accelerated from 0 to 100 km/h (62 mph) in

OPPOSITE: Resting on the platform of the Turbo, the GT3 RS showed its genes during competition.

LEFT: Sporty ambiance on board the GT3 RS. Cockpit ceiling, steering wheel and bucket seats in suede, lightweight door opening mechanism with straps and removal of the rear seats.

S GO 5011

3.3 seconds, compared to 3.5 seconds for the 991 GT3. This model met the same fate as its predecessor, which was produced in only 600 units. Although it was increased to 2,000 units, the production of the 2015 GT3 RS sold out even before the first deliveries. Thus, in 2018, the GT3 RS returned to the catalog in an even sharper style. The styling was largely inspired by the 911 GT2 RS, with wider fenders and a carbon fiber reinforced plastic front hood. Developed from the latest GT3's engine, the naturally aspirated 4-liter flat-six now delivered 520 hp. Compared to the 2015 version, the most significant improvements concerned the suspension and aerodynamics. The suspension arms were equipped with ball joints, instead of plastic bushings, and the springs were similar to those of the GT2 RS. The imposing rear wing generated 8% more downforce, and work on airflow increased high-speed stability. In the 0 to 100 km/h (62 mph) exercise, 0.1 seconds was gained. If the technical specifications mentioned a 10 kg (22 lbs) increase, it was due to the impossibility of homologating the car with the lightest equipment.

ENGINE: Flat-six
DISPLACEMENT: 4.0L / 243.9 in³
POWER: 500 hp at 8,250 rpm / 513 hp
TORQUE: 339 lb-ft at 6,250 rpm / 347 lb-ft at 6,000 rpm
GEARBOX: Six-speed manual or PDK seven-speed
DIMENSIONS (L x W x H): 14' 11.0" x 6' 2.0" x 4' 2.8"
WEIGHT: 3,131 lbs / 3,153 lbs
TOP SPEED: 193 mph / 194 mph

The 991-generation 911 GT3 RS returned in 2018 in a new, sharper version. The styling was largely inspired by the 911 GT2 RS, while the 4-liter flat-six now delivered 520 hp.

GT3 RS

911 R 2016

At the 2016 Geneva Motor Show, 911 enthusiasts rejoiced. Porsche presented a sports version that revived the spirit of the famous 1967 R, the first road-legal racing version of the 911. Developed from the 911 GT3 RS, the new R would never experience the hardships of racing, but it was intended to restore the pleasure of driving to its former glory.

From the original R, the 2016 R retained the lightweight philosophy. Compared to the GT3 RS, engineers managed to shed 50 kg (110 lbs) at 1,370 kg (3,020 lbs) by using magnesium for the roof and carbon for the hood and fenders. The rear window and side windows were made of polycarbonate. The spirit of the R permeated its clean body, free of its sister's ungainly adjustable rear wing. The reduction of this device resulted in less aerodynamic drag and thus an increase in top speed to 323 km/h (201 mph), a gain of 13 km/h (8 mph). However, efficiency and cornering stability required aerodynamicists to revise their design. While the front and rear bumpers were those of the GT3 RS, the R featured a front aerodynamic blade and a redesigned rear diffuser. These devices worked in conjunction with a retractable spoiler that differed from the Carrera's by a greater deployment. As for colors, nostalgia played a full role. White or gray, with or without stripes – red or green – and adorned with "Porsche" decorations on the side skirts, the 911 R perpetuated the spirit of its 1967 inspirer.

The 50 kg (110 lbs) weight saving also relied on reduced interior insulation, the removal of the rear seats in favor of a thin carpet, and a door opening system via a strap. The 2016 customer seemed no longer willing to forgo some comfort features. The center console and leather trim were retained, but Porsche offered purists the option of deselecting, in the free options column, air conditioning and the PCM infotainment screen including GPS navigation. Decidedly attentive to its customers, Porsche offered three types of seats, from Grand Touring bucket seats to racing shells.

ABOVE: A nod to history, the new 911 R adopted the side skirt decoration of the first 911 R from 1967.

RIGHT: By resurrecting the "R" designation associated with the first 911 competition car, Porsche triggered a real bidding war within the community of brand enthusiasts.

ABOVE: Benefiting from a limited production of 991 units, the 911 R was not the most powerful in the range, but the one that provided the most excitement.

RIGHT: While the interior of the 911 R was largely customizable, the standard upholstery revived the houndstooth pattern of the first 911s.

The 2016 911 R revived the original spirit of the 1967 R with its body stripped of a rear wing and its manual gearbox. It was also 50 kg (110 lbs) lighter than the GT3 RS from which it was derived.

If enthusiasts resorted to subterfuge to acquire an order form, it was as much because of the R's features as its numbered production limited to 991 units. The 2016 911 R was not intended to set lap records on the Nürburgring Nordschleife, which was confirmed by the return of the six-speed manual gearbox. It gave full flavor to driving by allowing a return to the delights of heel-and-toe. Engineers thought of those resistant to the exercise, and a Sport button activated automatic rev-matching, both on upshifts and downshifts. The gearbox paired perfectly with the latest evolution of the 4-liter naturally aspirated flat-six engine, producing 500 hp. Above 3,500 rpm, it dropped its mask to roar up to 8,500 rpm. While conceding a touch of efficiency and performance to the GT3 RS, the R still featured rear-wheel steering, Michelin Pilot Sport Cup 2 tires, and carbon-ceramic discs. The gratifying sensations it conveyed explained the collectors' avidity.

ENGINE: Flat-six
DISPLACEMENT: 4.0L / 243.9 in^3
POWER: 500 hp at 8,250 rpm
TORQUE: 339 lb-ft at 6,250 rpm
GEARBOX: Six-speed manual
DIMENSIONS (L x W x H): 14' 10.4" x 6' 0.9" x 4' 2.2"
WEIGHT: 3,020 lbs
TOP SPEED: 201 mph

PORSCHE
S GO 5023

911 1 MILLION 2017

This was undoubtedly the most important year in the life of the 911. After almost disappearing several times, the benchmark sports car crossed the one-million-unit production mark in May 2017. On this occasion, Porsche, always very attached to tradition, celebrated by manufacturing a unique model.

Named "911 1 Million," this model was based on a Carrera S equipped with the 3-liter twin-turbo flat-six producing 420 hp. Given the Porsche family's long-standing fondness for the color green, the car was finished in a bespoke "Iris Green" paint, complemented by "Gunmetal Grey" wheels, and metallic gray mirror housings and engine grille.

The interior received an extensive personalization treatment. In addition to unique instruments with chrome-ringed dials, the dashboard bears a plaque inscribed "911 Nr. 1,000,000." Like the earliest 911s, this milestone example reintroduces wood trim, featured on the steering wheel and as a strip running across the dashboard. The upholstery recalls the model's heritage, combining leather side bolsters with houndstooth fabric inserts reminiscent of many 1960s 911s.

A one-of-a-kind creation, the 911 1 Million was never offered for sale. Instead, it was reserved for display at the Porsche Museum, located near the company's historic headquarters and production facilities in Zuffenhausen.

In addition to the houndstooth fabric upholstery typical of the 1960s, this unique 911 features a dial with a stylized profile and seven-digit number.

ABOVE: Porsche waited fifty-four years to reach the millionth 911 produced. Intended for the brand's museum, this unique car featured the Porsche family's favorite color: a deep green called "Iris Green."

LEFT: One million 911s produced: this was celebrated by a unique version that rolled off the Zuffenhausen factory assembly line in May 2017 and featured specific instrumentation.

ENGINE: Flat-six biturbo
DISPLACEMENT: 3.0L / 181.9 in³
POWER: 414 hp at 6,500 rpm
TORQUE: 369 lb-ft from 1,700 to 5,000 rpm
GEARBOX: Seven-speed manual
DIMENSIONS (L x W x H): 14' 9.1" x 5' 11.2" x 4' 3.3"
WEIGHT: 3,175 lbs
TOP SPEED: 191 mph

911 GT3 TOURING PACKAGE
2017

The 911 GT3 Touring Package was the consolation prize intended to alleviate the frustration of customers who were turned away from the 911 R. Presented in September 2017, the new GT3 variant had everything of an R, except for its limited production. Reviving the "Touring package" designation of the 911 2.7 Carrera RS, this GT3 variant was distinguished by a rear hood with a fixed spoiler, rather than the retractable appendix of the 911 Carrera. A GT3 Touring logo was affixed to the grille. The Grand Touring-focused ambiance abandoned Alcantara for smooth leather. The positioning of the GT3 Touring package was reflected in the singular choice of the six-speed manual gearbox. Like that of the 911 R, the transmission featured automatic rev-matching.

Under the hood, nothing changed compared to the GT3. Coming from the 911 RSR and GT3 Cup, the naturally aspirated 4-liter flat-six engine still developed 500 hp and remained capable of reaching 9,000 rpm on intermediate gears. Significantly less refined than the 911 R, the GT3 Touring Package was credited with a top speed of 316 km/h (196 mph) instead of 323 km/h (201 mph). It also conceded a small tenth of a second on the 0 to 100 km/h (62 mph) sprint, which it completed in 3.9 seconds. With identical running gear and tire setup, the difference with the R remained insignificant on the road.

BELOW: As with the 2016 911 R, the naturally aspirated 4-liter flat-six engine produced 500 hp, and the gearbox was a six-speed manual.

ENGINE: Flat-six
DISPLACEMENT: 4.0L / 243.9 in³
POWER: 500 hp at 8,250 rpm
TORQUE: 339 lb-ft at 6,250 rpm
GEARBOX: Six-speed manual
DIMENSIONS (L x W x H): 14' 11.6" x 6' 0.9" x 4' 2.1"
WEIGHT: 3,115 lbs
TOP SPEED: 196 mph

ABOVE: Porsche consoled the unsuccessful applicants for the 911 R with the GT3 Touring derivative. Its features were similar, but the new car was not produced in a limited series.

The 911 GT3 Touring could be fitted with the optional Clubsport package, which was indicated by a stopwatch located at the top of the dashboard.

911 GT2 RS
2017

ENGINE: Flat-six biturbo
DISPLACEMENT: 3.8L / 231.9 in^3
POWER: 690 hp at 7,000 rpm
TORQUE: 553 lb-ft from 2,500 to 4,500 rpm
GEARBOX: PDK seven-speed
DIMENSIONS (L x W x H): 14' 11.1" x 6' 2.0" x 4' 3.1"
WEIGHT: 3,241 lbs
TOP SPEED: 211 mph

As the 991 generation approached the end of its life, it still had not spawned a GT2 variant. That absence was remedied in June 2017 at the Goodwood Festival of Speed in England with a new 911 GT2 RS. This time, the limits seemed to have been reached. With the 3.8-liter twin-turbo flat-six with 700 hp transmitted to the rear wheels to move a mass of less than 1.5 tons, Porsche had never served such an explosive cocktail. Becoming the most powerful road-legal version of all time from the Zuffenhausen firm, the GT2 RS set a new record on the Nürburgring Nordschleife in 6 minutes 47.3 seconds. That was ten seconds faster than the famous 918 Spyder. The 100 km/h (62 mph) sprint was devoured in 2.8 seconds, the 200 km/h (124 mph) in 8.3 seconds. The thrust seemed never-ending up to the maximum speed of 340 km/h (211 mph), where downforce reached 416 kg (917 lbs).

The engineers in Weissach were not content with increasing power by installing large turbochargers and water cooling for the supercharging system. The 911 GT2 RS benefited from lessons learned in competition and from the 911 GT3 RS in terms of aerodynamics and chassis dynamics. With the louvered vents on top of the front fenders, air intakes in the rear fenders, a front bumper with two enormous square air intakes, NACA ducts in the front hood, and a two-position adjustable rear wing, the GT2 RS inherited the full array of solutions used in competition.

The GT2 RS's efficiency relied on four-wheel steering, a revised PSM system, specific Michelin Pilot Sport Cup 2 tires mounted on the widest wheels Porsche had ever put on a 911 – 265/35 ZR 20 at the front and 325/30 ZR 21 at the rear – and large-diameter carbon-ceramic discs.

As was often the case with Porsche, the GT2 RS proved perfectly compatible with daily use. The removal of soundproofing material resulted in a high noise level in the cabin and the transmission of road imperfections.

BELOW: The 911 GT2 RS benefited from lessons learned from motorsport in terms of aerodynamics and weight reduction.

ABOVE: Thanks to carbon fiber-reinforced plastic (CFRP) body parts and a magnesium roof, the weight did not exceed 1,470 kg (3,241 lbs). 30 kg (66 lbs) could be further saved by opting for the optional Weissach package.

ABOVE: The GT2 RS arrived only at the end of the seventh generation's life, but it was worth the wait. This model became, in 2017, the most powerful and highest-performing road-legal 911 ever produced.

911 TURBO S EXCLUSIVE SERIES 2017

Appearing in 2017, the 911 Turbo S developed by Porsche's "Exclusive Manufaktur" workshop featured central gold stripes on its leather upholstery and stitching of the same color.

In charge of the most extensive customizations, the Porsche "Exclusive" workshop, renamed "Exclusive Manufaktur," developed a very special version of the 911 Turbo S during the first half of 2017. Produced in a limited series of 500 units under the name "Exclusive Series," this model was distinguished by its metallic gold color, coordinated with 20-inch black center-lock wheels, carbon-ceramic brake calipers in black, and two woven carbon stripes running through the center of the car. The hood, side skirts, active rear wing, and roof were made of carbon.

The bespoke character of this Turbo S was accentuated by the presence of the rear wing from the aerodynamic kit, a new rear apron, a carbon air intake manifold, a dual exhaust system with two black steel outlets, and Porsche "Exclusive Manufaktur" plaques placed on the front fenders. The interior treatment reflected the exterior decoration. The leather sport seats featured two central gold stripes. This color was also found on the stitching, the Turbo S signature embroidered on the headrests, and the stripes adorning the headliner. A plaque indicating the series number and the illuminated "Exclusive Series" monogram on the door sills confirmed the unique character of the model.

The engine further distinguished this special series from the models in the range. The 3.8-liter twin-turbo flat-six gained 27 hp to reach 607 hp making it the most powerful 911 Turbo in history. It reached 100 km/h (62 mph) in 2.9 seconds, and its top speed was set at 330 km/h (205 mph). It was based on a sport chassis associated with PASM adaptive suspension, four-wheel steering, and the Porsche Dynamic Chassis Control (PDCC) anti-roll system. Porsche Design also produced a chronograph matching the vehicle, limited to 500 examples.

Porsche produced only 500 numbered examples of its 911 Turbo S Exclusive Series, which was recognizable by its metallic gold paint and its aerodynamic kit.

ENGINE: Flat-six biturbo
DISPLACEMENT: 3.8L / 231.9 in^3
POWER: 599 hp at 6,750 rpm
TORQUE: 553 lb-ft from 2,250 to 4,000 rpm
GEARBOX: PDK seven-speed
DIMENSIONS (L x W x H): 14' 9.4" x 6' 2.0" x 4' 3.1"
WEIGHT: 3,527 lbs
TOP SPEED: 205 mph

The 911 Turbo S Exclusive Series was not just a refined finish. It became the most powerful 911 Turbo in the catalog with its 607 hp 3.8-liter twin-turbo flat-six.

911

935
2018

The secret was well-kept, so the spectators of the "RennSport Reunion" in September 2019, a gathering of German historical race cars on the Laguna Seca circuit in California, were far from suspecting that they would witness the world premiere of a model undoubtedly resulting from the craziest project of the 2010s. The era was about resurrecting legends of the past. Designer Grant Larson and his team rode this trend by giving a successor to the famous 935/78 Moby Dick, which raced at the 24 Hours of Le Mans in 1978 with Manfred Schurti and Rolf Stommelen. As always with this type of project, the commando had only four days to complete the exterior design based on a 911 GT2 RS. A truly special program: "We had total freedom since there were no homologation constraints. The engineers were able to give free rein to their ideas," declared Larson, the head of development for the new 935.

Porsche designed this car with as much care as if it were a competition vehicle entered in the biggest sporting events. Thus, the artisans of the project relied on recent progress made in all areas to develop a car that embodied the latest technology. With its aerodynamic body evoking the 1978 model, the length reached 4.85 m and the width 2.03 m. The use of carbon fiber reinforced composite body elements kept the weight at 1,380 kg (3,042 lbs). Based on the 911 GT2 RS, the 2018 935 was not homologated for road or race, but designed to participate in brand sporting events and private track tests. The 700 hp 3.8-liter twin-turbo flat-six was mated to the PDK seven-speed dual-clutch transmission.

The cockpit resembled that of the 911 GT2 RS Clubsport, but some details referred to the 1970s, such as the laminated wood gear lever. Only 77 examples of this car, which could be extensively customized, were produced.

The stripped-down interior of the 935 was that of a competition car with its digital information screen, enveloping bucket seat, and composite doors.

ENGINE: Flat-six biturbo
DISPLACEMENT: 3.8L / 231.9 in³
POWER: 690 hp at 7,200 rpm
TORQUE: 553 lb-ft from 2,500 to 4,500 rpm
GEARBOX: PDK seven-speed
DIMENSIONS (L x W x H): 15' 11.5" x 6' 8.1" x 4' 5.2"
WEIGHT: 3,042 lbs
TOP SPEED: 211 mph

RIGHT: Porsche was inspired by the 935 with Moby Dick bodywork that finished eighth in the 1978 24 Hours of Le Mans to market a series of 77 cars, based on the 911 GT2 RS Clubsport.

BELOW: This was Porsche's craziest project of the late 2010s. Freed from all homologation constraints, the engineers created an impressive, true racing car.

911 SPEEDSTER
2018

What car best embodied the Porsche spirit? Without a doubt, the Speedster, which harks back to the company's origins with its shortened windshield on a roadster base. Thus, in its seventieth year, the German manufacturer reintroduced this formula based on the 991 generation. Unveiled as a concept in the summer of 2018, the production Speedster debuted at the Paris Motor Show in October 2018. This new model also celebrated the thirtieth anniversary of the first 911 of its kind. Once again, the brand risked incurring the wrath of its customers, as it decided not to produce more examples than its birth year.

Produced in only 1,948 units, the Speedster closed the 991 type series. It combined a widened 911 Carrera 4 body with the chassis and 4-liter engine of the 911 GT3. A large part of the appeal of this strict two-seater lay in its carbon fiber double-bubble rear hood, its shortened windshield, and reduced side windows, necessitating a very low and forward-sloping soft top.

The technical manual emphasized that the soft top was only for emergency use which implied that the Speedster was not a cabriolet and should be driven with the wind in one's hair as often as possible. The entirely manual soft top followed a precise choreography. Only the unlocking of the hooks adjusted on the windshield frame and the rear cover benefited from electric assistance.

The Heritage Design package, based on a cognac leather interior and an exterior decoration of metallic silver paint, reinforced the Speedster's personality. Under the hood of the 2018 Speedster, the naturally aspirated 4-liter flat-six delivered 510 hp at 8,400 rpm. It was mated to the six-speed manual gearbox to provide old-school driving pleasure.

Produced in only 1,948 units, in reference to the year of the brand's birth, the Speedster closed the 991 generation and announced the new Heritage Design family.

Dedicated to sensation, the 991 Speedster was a delightful cocktail that combined a widened 911 Carrera 4 body with the chassis and 4-liter engine of the 911 GT3.

ENGINE: Flat-six
DISPLACEMENT: 4.0L / 243.9 in^3
POWER: 503 hp at 8,400 rpm
TORQUE: 347 lb-ft at 6,250 rpm
GEARBOX: Six-speed manual
DIMENSIONS (L x W x H): 14' 10.4" x 6' 0.9" x 4' 1.2"
WEIGHT: 3,230 lbs
TOP SPEED: 193 mph

The interior environment of the 991 Speedster was largely inspired by that of a GT3 Touring. As with previous Speedsters, the soft top was manual.

911 GT2 RS CLUBSPORT 2018

The competition version of the 911 GT2 RS, the Clubsport version, immersed the driver in an RSR environment with a simplified steering wheel and reinforcements in the doors.

The prospect of participating in major Grand Touring category races with chances of victory and offering a high-performance vehicle to enthusiasts led Porsche to design a competition version of the 911 GT2 RS type 991.2. Presented at the Los Angeles Auto Show in November 2018, the GT2 RS Clubsport was produced in a limited series of 200 examples. Like the 935, it was powered by a 3.8-liter twin-turbo flat-six, with four valves per cylinder, variable camshaft timing, and Variocam Plus intake valve lift adjustment. The 700 hp power was transmitted to the 310 mm (12.2 in) wide rear wheels via a PDK dual-clutch gearbox with seven ratios controlled by paddle shifters on the steering wheel. To stop this high-performance car, engineers installed six-piston aluminum monobloc racing calipers and 390 mm (15.34 in) diameter grooved and ventilated steel discs on the front axle. At the rear, the braking system relied on 380 mm (14.96 in) discs and four-piston calipers.

Like the GT2 RS, the Clubsport paid attention to its lines. It weighed no more than 1,390 kg (3,064 lbs) with a body combining aluminum elements, an imposing rear wing, and a carbon fiber reinforced plastic roof. The interior ambiance was that of a racing car with an enveloping bucket seat equipped with a six-point safety harness, a welded roll cage, and racing instrumentation borrowed from the GT3 R. Air conditioning was retained to prevent the cabin from turning into a hotbox and increasing driver fatigue. The GT2 RS Clubsport was also equipped with the PSM system with traction control as well as ABS braking system. These aids could be adjusted separately or completely deactivated.

In the summer of 2021, Porsche celebrated 25 years of its partnership with the Manthey racing team by marketing 30 units of the GT2 RS Clubsport.

Presented at the Los Angeles Auto Show in October 2018, the 911 GT2 RS Clubsport shared many elements with the 935.

The close ties between Porsche and the Manthey racing team, founded by driver Olaf Manthey, led Porsche to celebrate the relationship's twenty-fifth anniversary with an incredible competition machine produced in a limited series of thirty examples. Based on the 911 GT2 RS Clubsport, the new car was unveiled in the summer of 2021. The technology was similar to that of the 2018 car, but Porsche Motorsport technicians went even further in seeking aerodynamic efficiency. The front apron was entirely redesigned around an enormous central air intake and two lateral ducts. NACA ducts were integrated into the front hood, while the hypertrophied rear fenders encircled the rear end. The modifications continued with a rear wing with swan-neck supports and new side plates, as well as ventilation openings in the engine hood under the rear window. The two exhaust pipes exited at the center of the bumper.

The position of the central radiator was one of the Clubsport's essential characteristics. This concept eliminated the radiators in the wheel arches. As for the suspension, it took the main components from the 935, but the offsets were greater. Porsche added anti-rain taillights from the 911 GT3 Cup type 992 and four-point headlights inherited from the 911 RSR.

ENGINE: Flat-six biturbo
DISPLACEMENT: 3.8L / 231.9 in^3
POWER: 690 hp at 7,000 rpm
TORQUE: 553 lb-ft from 2,500 to 4,500 rpm
GEARBOX: PDK seven-speed
DIMENSIONS (L x W x H): 15' 6.7" x 6' 5.9" x 4' 5.5"
WEIGHT: 3,064 lbs
TOP SPEED: N/A

PORSCHE
S GO 4039

VIII

TYPE 992

—

since 2019

With the eighth generation, the 911 truly entered its century, converting to new technologies, for the benefit of pleasure, safety and performance. Versatility, at the heart of its design, supports the irreconcilable challenge: winning new customers, mainly in expanding markets, without losing the brand's loyal customers. During its career, the 992 set new benchmarks in terms of technology by introducing a new hybrid powertrain.

The transition to the eighth generation once again became a high-risk operation. Where to set the cursor between reasoned evolution and revolution, so as not to disappoint enthusiasts and conquer new customers? Even if the brand assumed a certain form of continuity, the type 992, presented at the Los Angeles Auto Show at the end of 2018, evolved much more than its appearance suggested. A few precise strokes were enough to modernize the inimitable and unchanging line of the 911.

While the wheelbase was identical to that of the 991, the length gained 20 mm (.78 in) and, above all, the tracks widened + 45 mm (1.77 in) at the front. Add 20-inch wheels at the front and 21-inch wheels at the rear, and you had a more voluminous car that appeared ready to pounce. The rear fenders of the 991 GTS were identical for the Carrera S and 4S, making it impossible to distinguish them.

Aside from bumpers covered in black plastic, the most significant changes concerned the integration of retractable door handles and the treatment of the rear. With its illuminated strip connecting the slim taillights and the massive bumper, the rear recalled the brand's SUVs. The 80% redesigned platform opened access to new technologies. This allowed it to limit the weight gain to 50 kg (110 lbs) by incorporating a 12 kg (26 lbs) reduction in body weight by using nearly 70% aluminum. The interior ambiance approached that of the brand's sedans, emphasizing status rather than sportiness. The instrumentation was housed in a curved slab. The essential was preserved. In the middle of the two customizable screens sat an enormous analog tachometer. The center of the dashboard strip was occupied by a 10.9-inch touchscreen. It eliminated an army of buttons, without degrading ergonomics. Always intended to be easier to live with in order to conquer new customers, the 911 type 992 multiplied its driving aids.

The only model available at the launch in March 2019, the Carrera S remained faithful to the 3-liter twin-turbo flat-six. Equipped with a particulate filter for depollution and piezoelectric injectors for combustion, it gained 30 hp to reach 450 hp at 6,500 rpm and 530 Nm (391 lb-ft) of torque from 2,300 to 5,000 rpm. It was mated to a new eight-speed PDK dual-clutch gearbox and could access all-wheel drive. Beyond superior performance – a 0.4-second gain to reach 100 km/h (62 mph) in 5 seconds on the Nürburgring Nordschleife – the new 911 appeared easier to drive. It introduced a "wet" mode that allowed driving on wet roads without apprehension. This system detected humidity using sensors in the wheels and adapted the vehicle's components accordingly. The Carrera version followed a few months later. With the 2,981 cm3 flat-six equipped with smaller turbos, power dropped to 385 hp and torque to 450 Nm (332 lb-ft). The 100 km/h (62 mph) sprint was achieved in 4.2 seconds. The Carrera did not have rear-wheel steering, an option on the S, while the braking system was reduced in size.

Versatility, at the heart of its design, supported the challenge of conquering new customers, mainly in expanding markets, without losing loyal brand enthusiasts. During its career, the

RIGHT: Eighty percent redesigned from the model it replaced, the 992 featured a new dashboard and gave access to the latest cutting-edge technologies.

PREVIOUS TWO PAGES: With the eighth generation, the 911 truly entered its century, converting to new technologies, benefiting driving pleasure, safety, and performance.

992 set new benchmarks in technology by debuting a new hybrid powertrain.

At Porsche, there was no question of altering the usual production processes, and the different body variants gradually enriched the 911 range. Thus, the Cabriolet appeared, successively followed by the 911 Turbo, the Targa, and the Carrera GTS. Based on the 992, the new Turbo was to be presented at the Geneva Motor Show in 2020, which was canceled at the last minute due to the coronavirus health crisis. The body was widened by 45 mm (1.78 in) at the front to 1,840 mm (72 in) and by 20 mm (.78 in) at the rear to reach 1.90 m (75 in). At the same time, the front track was widened by 42 mm (1.65 in) while the rear track extended by 10 mm (.39 in). The rear spoiler followed the same movement, taking up even more space and adopting an airbrake function. An active aerodynamic system was almost indispensable to stop this model capable of going from 80 to 120 km/h (75 mph) in 1.6 seconds.

The 3.8-liter engine with two variable geometry turbos gained 70 hp to deliver 650 hp and 800 Nm (590 lb-ft) of torque. For the first time, the 911 Turbo S was equipped with different tire sizes: 20-inch wheels at the front, 21-inch at the rear. A detuned version with 580 hp followed a few months later. Meanwhile, the 911 Carrera GTS's power increased to 480 hp and its torque to 570 Nm (420 lb-ft). The chassis, lowered by 10 mm (.39 in), gained access to PASM adaptive suspension. For the first time, the GTS could receive a Lightweight package reducing weight by 25 kg (55 lbs) and integrating rear-wheel steering. The 911 seemed to be endlessly diversified. At the end of 2022, it was enriched with a T variant based on a 911 Carrera lightened by 35 kg (77 lbs) and equipped with the Carrera S chassis.

Though anticipated for some time, the electrification of the 911 Carrera became effective in the summer of 2024. The benchmark sports car thus crossed a stage as important as the transition to water cooling in 1996 and the adoption of supercharging in 2015. While the S version disappeared from the catalog, the GTS model was new, featuring T-Hybrid technology. The 3-liter twin-turbo flat-six gave way to a 3.6-liter single-turbo engine with 15 hp electric assistance. With 485 hp, this engine combined with a 54 hp permanent magnet synchronous electric motor installed at the end of the crankshaft and powered by a 1.9 kWh battery placed under the front trunk. The combined power was 541 hp, and torque was 650 Nm (479 lb-ft). Performance improved significantly with a 0 to 100 km/h (62 mph) time of 3 seconds, a top speed of 312 km/h (194 mph), and 8.7 seconds gained at the Nürburgring. These innovations were accompanied by a front bumper with ten active air flaps working with adaptive diffusers and a fully faired underbody, redesigned headlights, a white rear strip, and two exhaust pipes in the center of the bumper. The GTS's efficiency relied on rear-wheel steering and the PDCC roll management system. The rear seats were a free option. Furthermore, the 911 Carrera continued its career: the 3-liter twin-turbo gained 9 hp to deliver 394 hp. From autumn 2024, it equipped the Carrera T, now available with a cabriolet body.

911 TARGA 4S HERITAGE DESIGN EDITION 2020

Porsche was one of the rare manufacturers able to afford a formula of marrying present and past. This materialized in early June 2020 with the birth of a new limited-series 911 family. Under the Heritage Design signature, Porsche entrusted its customization department Porsche "Exclusive Manufaktur" with the creation of a series of four models that successively reinterpreted the 1950s, 1960s, 1970s, and 1980s.

The first of the lineage was based on the 911 Targa 4S, from which designers imagined a decoration and finish that recalled the features popular in the 1950s and 1960s. The Porsche Heritage badge – affixed to the grille – referred to the one formerly issued to owners of Porsche 356s who had crossed the 100,000 km mark. Another nod to the past was the Porsche crest, adorning the front hood, steering wheel, wheel covers, and ignition key, was that of 1963. The four available body colors, including the exclusive Metallic Cherry, were accompanied by a decoration often present on racing models. The 20 and 21-inch wheels with black-painted brake calipers also revived the five-spoke design popular in the 1960s.

The interior treatment was also a tribute to history. Designers developed two interior ambiances coordinated with the chosen body color: burgundy or black combined with Atacama beige.

ENGINE: Flat-six biturbo
DISPLACEMENT: 3.0L / 181.9 in³
POWER: 444 hp at 6,500 rpm
TORQUE: 391 lb-ft from 2,300 to 5,000 rpm
GEARBOX: PDK eight-speed
DIMENSIONS (L x W x H): 14' 9.9" x 6' 0.9" x 4' 3.2"
WEIGHT: 3,616 lbs
TOP SPEED: 189 mph

The seat upholstery was inspired by that of the 356, combining corduroy and leather. The lighting or green lettering of the tachometer and chronograph, as well as the numbered plaque fixed on the dashboard, added the final touch to this nostalgic journey. Especially for this model, Porsche Design produced a mechanical chronograph evoking the tachometer of 1950s Porsches with a white second hand and green rings.

Produced in only 992 examples in reference to the generation from which the first Heritage Design model was conceived, this Targa 4S shared its technical characteristics with the 911 Carrera 4S.

BELOW: In the lineage of the famous Art Cars, American artist Daniel Arsham designed the livery of Arthur Mathieu's GT3 Cup.

ABOVE: Régis Mathieu's son, Arthur Mathieu, a passionate collector of the German brand, was one of the youngest participants in the French Carrera Cup.

RIGHT: The sixth generation of the 911 Carrera Cup, the 992 generation, paid attention to mass distribution, lowering the center of gravity, and aerodynamics.

ENGINE: Flat-six
DISPLACEMENT: 4.0L / 243.9 in^3
POWER: 503 hp at 8,400 rpm
TORQUE: 347 lb-ft at 6,150 rpm
GEARBOX: Six-speed sequential
DIMENSIONS (L x W x H): 15' 0.5" x 6' 3.6" x 4' 5.2"
WEIGHT: 2,778 lbs
TOP SPEED: N/A

911 GT3 CUP
2021

When the Porsche Motorsport department presented the new GT3 Cup developed from the 992, 4,251 units of the 911 Cup had already been produced in thirty years. With the new GT3 Cup, intended to enter service in the various championships from the 2021 season, Porsche now aimed for 5,000 units. This new model benefited greatly from the experience gained with the preceding model, but also from the 911 RSRs competing in endurance races. The engineers at the Weissach center started work in early 2019. Under project manager Jan Feldmann, the team focused on improving performance while ensuring faster interventions, greater component longevity, and reduced maintenance costs.

For the first time in the history of the Carrera Cup, the GT3 Cup was based on a lightweight Turbo body. Sitting on 12-inch wide front wheels and 13-inch wide rear wheels that hugged the wide fenders, the race car was impressive. With a total width of 1,902 mm (74.9 in), the body was widened by 28 mm (1.1 in) at the rear compared to the previous model and was distinguished by its additional cooling air intakes in front of the wheels. While the 991/2 body was composed of 70% steel and 30% aluminum, the ratio was reversed. However, the GT3 Cup gained 35 kg (77 lbs) to 1,260 kg (2,778 lbs) because of additional spacers in the safety cell. The doors, engine hood, and rear wing were made of carbon fiber reinforced plastic, while the front hood remained aluminum.

The aerodynamic concept was extensively revised to optimize downforce. It relied primarily on the front apron combined with two rear wings: a fixed one similar to that of a Turbo, topped by another device with a swan-neck support adjustable in eleven positions.

Finally, Porsche improved the ergonomics of the 992 GT3 Cup's cockpit, whose naturally aspirated 4-liter flat-six engine delivered 510 hp, a gain of 25 hp compared to the model it replaced.

LEFT: In the lineage of the famous Art Cars, American artist Daniel Arsham designed the decoration of Arthur Mathieu's GT3 Cup.

OPPOSITE: The sixth generation of the 911 Carrera Cup, the 992, emphasized mass distribution, lowering the center of gravity, and aerodynamics.

PORSCHE
TAGHeuer
A. Mathieu
53
www.porsche.fr
Pierre Martinet
MICHELIN
TAGHeuer
Centres Porsche Fréjus etToulon

911 GT3
2021

ENGINE: Flat-six
DISPLACEMENT: 4.0L / 243.9 in³
POWER: 503 hp at 8,400 rpm
TORQUE: 347 lb-ft at 6,100 rpm / 332 lb-ft
GEARBOX: PDK seven-speed / six-speed manual
DIMENSIONS (L x W x H): 15' 0.0" x 6' 0.9" x 4' 2.4"
WEIGHT: 3,164 lbs
TOP SPEED: 193 mph / 194 mph

One figure was enough to summarize the progress made by the 911 GT3 since its appearance with the 996 generation: just 6 minutes 59.927 seconds to complete a lap of the Nürburgring Nordschleife, a saving of one minute in twenty-two years. This result was attributed to a widened body housing enormous 21-inch wheels, perfected aerodynamics relying on vents in the front hood, a redesigned rear diffuser, a swan-neck mounted rear wing with four adjustment positions, and revisited 992 axles.

The interaction of the different devices considerably improved downforce compared to the previous model. The GT3 benefited greatly from lessons learned in competition. The front axle thus received the double wishbone front suspension from the 911 RSR, providing superior stability in corners and during braking. These devices combined with a 20 mm (.79 in) lowered chassis, partially deactivatable stability control (PSM), rear-wheel steering – up to 2 degrees of steering angle – and larger diameter brake discs 408 mm (16.06 in).

Like all members of the GT3 lineage, the 992 version was careful with its design. Thanks to a carbon fiber reinforced plastic (CFRP) front hood, a lightweight rear hood, lightweight glass windows, forged iron wheels, and a 10 kg (22 lbs) lighter stainless steel sport exhaust, the fourth generation of the sporty 911 weighed only 5 kg (11 lbs) more than the previous model. The weight was set at 1,435 kg (3,164 lbs) with the PDK seven-speed automatic transmission, which was 17 kg (37 lbs) more than with the six-speed manual gearbox. Coming from the 911 GT3 Cup, the naturally aspirated 4-liter engine gained 10 hp to reach 510 hp. The standard equipment finally included the "Track Screen" function, which reduced the display to the information necessary for track driving.

Following the launch of the revised 911 Carrera in July 2024, Porsche introduced the GT3. This car was unveiled on October 18 at the Hockenheim circuit in Germany, on the sidelines of the DTM championship final. On the engine side, there was no change. The engineers focused on complying with increasingly draconian emission standards. At the cost of adopting two particulate filters and four catalytic converters, the 4-liter flat-six succeeded in conforming.

At the DTM final at Hockenheim in October 2024, Porsche presented an evolution of its GT3, distinguished by the adoption of two particle filters and four catalysts to comply with pollution control standards.

With its 992 GT3, Porsche refined the supersports car concept. This coupé benefited more than ever from the lessons learned in competition, and notably from the RSR, from which it inherited the superimposed wishbones.

The engineers improved high-rev operation with new cylinder heads and optimized camshafts. The engine was also equipped with optimized flow throttle bodies and an improved oil cooling system. Power did not change, but both transmissions received an 8% shorter gear ratio. The PDK gearbox allowed a 0 to 100 km/h (62 mph) time of 3.4 seconds, compared to 3.9 seconds for the manual gearbox. Lacking increased power, the GT3 was lighter. Silver aluminum wheels reduced unsprung mass by more than 1.5 kg (3.3 lbs) compared to the previous version. Magnesium wheels were also offered as an option with the Weissach package or with the lightweight design package, saving 9 kg (20 lbs). The weight was set at 1,420 kg (3,131 lbs) in its lightest configuration. Another innovation was bucket seats with a thoracic airbag and headrest padding that could be removed to facilitate helmet wearing. For the first time, the upper part of the dashboard was made of a material imitating Alcantara. The dashboard instrumentation became digital. The GT3's efficiency was reinforced by the GT3 RS's front axle, an anti-dive system, and new curved sideplates for the rear wing. The Weissach package added a stabilizer bar and a carbon reinforcement plate on the rear axle.

ABOVE: Despite larger wheels and a wider body compared to the 991, the 992 GT3 limited its weight gain to only 5 kg (11 lbs).

RIGHT: The 992 GT3 emphasized efficiency and driving comfort. A new function called "Track Screen" adapted the screen display for track use.

S GT 3992

OPPOSITE: Since the version presented in October 2024 at the Porsche Experience Center in Hockenheim, the GT3 Touring Package could be fitted with rear seats as a free option.

911 GT3 TOURING PACKAGE
2021

In parallel with the 911 GT3, Porsche reintroduced the Touring variant. The two models shared the main technical characteristics, but the Touring version distinguished itself from its sister with side windows surrounded by an aluminum trim and by the absence of the fixed rear wing, in favor of a retractable and automatically deploying rear spoiler. Similarly, the front was entirely painted in the body color. With the Touring package, the interior ambiance emphasized comfort and refinement with numerous leather elements: steering wheel rim, gear lever, storage compartment cover on the center console, armrests in the door trims, and door handles. The front of the dashboard and the upper part of the side trim of the doors were distinguished by a special embossing. Equipped as standard with the six-speed manual gearbox allowing automatic rev-matching on downshifts, the GT3 Touring Package could also, for the first time, be ordered with the PDK seven-speed dual-clutch automatic transmission, at no extra cost.

Both the GT3 and the Touring were improved in autumn 2024. On both versions of the GT3, the front diffuser, front apron, and aerodynamic profile of the side skirt were redesigned to increase downforce and optimize airflow. New standard matrix LED headlights and optional HD Matrix LED headlights provided all the lighting functions of the 911, making additional lights in the front bumper superfluous. As a result, the bumper could

BELOW: Slightly less powerful than the GT3, the GT3 Touring Package was primarily aimed at those who preferred a GT3 without a rear wing and was less ostentatious.

accommodate enlarged air intakes. The 2024 Touring Package still dispensed with a rear wing, but at customer request, rear seats were made available. To ensure optimal access, the front bucket seatbacks became foldable. This GT3 variant could access the Leichtbau package, which featured the same setup as the Weissach package, with the only difference being the carbon hood and roof painted in body color.

ENGINE: Flat-six
DISPLACEMENT: 4.0L / 243.9 in^3
POWER: 503 hp at 8,400 rpm
TORQUE: 347 lb-ft at 6,100 rpm / 332 lb-ft
GEARBOX: PDK seven-speed / six-speed manual
DIMENSIONS (L x W x H): 14' 11.9" x 6' 0.9" x 4' 2.4"
WEIGHT: 3,126 lbs
TOP SPEED: 199 mph / 193 mph

The "Drive" rotary knob on the right side of the steering wheel managed the various driving programs, including "Race" mode, which deactivated ESP and traction control.

911 SPORT CLASSIC 2022

The second model in the Heritage Design series was revealed on April 28, 2022. Evoking the 1960s and 1970s, it revived the name and spirit of the 911 Sport Classic that appeared in 2009. Similarities were limited to the symbolic ducktail and the Sport Gray color – black, Quartz Gray Metallic, Gentian Blue Metallic, and special colors from the "Paint to Sample" customization program were also available. The rest was entirely new. Purists saw this as a departure from history. Tasked with celebrating the fiftieth anniversary of the famous 911 2.7 Carrera RS, the new Sport Classic was not developed from a 911 GT3 but from a 911 Turbo 992.

Like the Targa 4S from the same lineage, the 911 Sport Classic featured white-backed racing roundels on the doors, with light Sport Gray stripes running through the center of the body. The heritage of the 1960s and 1970s is still evident, with Fuchs-style wheels – here featuring center locks – a double-bubble roof, and rear fenders devoid of air intakes. The spirit of the 1960s and 1970s was fully respected with a rear-wheel-drive transmitting power only to the rear wheels and a seven-speed manual gearbox. While contributing to the model's character, the transmission limited torque to 600 Nm (443 lb-ft) and power to 550 hp.

Prioritizing driving sensations over pure performance, the 911 Sport Classic was distinguished by lower aerodynamic downforce than the production Turbo. In return, this limited-edition model lost 70 kg (154 lbs) to 1,570 kg (3,461 lbs) by using composite materials – plastic coupled with carbon fiber – for the hood, roof, and rear spoiler. From the first generation of the 911, the 2022 Sport Classic revived the legendary Pepita houndstooth fabric with cognac-colored leather to cover the door panels and the central stripe of the seats.

Produced in only 1,250 units, the 911 Sport Classic included, as standard equipment, carbon-ceramic brake discs, PASM suspension lowered by 10 mm (.39 in), and active chassis control PDCC. Under the rear hood with the ducktail of the famous 2.7 Carrera RS, the 3.7-liter twin-turbo flat-six from the 911 Turbo breathed. It was, however, detuned by 30 hp due to the presence of a seven-speed manual gearbox.

Under the rear hood equipped with the ducktail of the famous 2.7 Carrera RS breathed the 3.7 liter twin-turbo flat-six of the 911 Turbo. However, it is down-tuned by 30 hp due to the presence of a seven-speed manual transmission.

ABOVE: Rear-wheel drive, 10 mm (.39 in) lowered chassis, manual transmission: the 911 Sport Classic revived old-school driving pleasure.

LEFT: The formula was well-established. The second model in the Heritage Design series featured the famous Pepita houndstooth fabric upholstery and the green instrumentation of 1960s models.

ENGINE: Flat-six biturbo
DISPLACEMENT: 3.7L / 228.5 in³
POWER: 543 hp at 6,750 rpm
TORQUE: 443 lb-ft from 2,000 to 6,000 rpm
GEARBOX: Seven-speed manual
DIMENSIONS (L x W x H): 14' 10.5" x 6' 2.8" x 4' 3.1"
WEIGHT: 3,461 lbs
TOP SPEED: 196 mph

911 GT3 RS
2022

According to the German manufacturer, an increasing number of sports car enthusiasts were interested in driving on tracks without racing. For them, Porsche created a new generation of the 911 GT3 RS. This car was designed to be driven on the road to a circuit, challenge the stopwatch, offer thrills and performance close to a race car, and then return home by road.

Developed from a GT3, the GT3 RS strikingly resembled the 911 RSRs that competed in top endurance events. Never had Weissach engineers pushed the transfer of technology between competition and road so far. Based on a Turbo body with tracks widened by 28 mm (1.1 in) at the front, this new model seemed to be a technical replica of the Le Mans 911s in its aerodynamics, chassis, and engine.

OPPOSITE TOP: As usual at Porsche, almost anything was possible in terms of interior trim. A Weissach package, featuring a roll cage made of carbon-fiber-reinforced plastic (CFRP) and ultra-thin bucket seats, helps shave off a few extra kilos.

ABOVE (BOTH): With the 992 GT3 RS, a production Porsche had never gone so far in transferring race technology to the road. It featured a DRS (Drag Reduction System) based on winglets and vents.

The aerodynamic downforce approached that of a racing 911 at 409 kg (902 lbs) at 200 km/h (124 mph) and 860 kg (1,896 lbs) at 285 km/h (177 mph). This performance relied on a combination of solutions including an aerodynamic lip at the base of the bumper, vents in the front fenders, vertical winglets around the wheel arches, deflectors on the front hood air intakes, slats on the roof, active flaps under the front bumper, flanks that curved near the wheels, a fully faired underbody, profiled front suspension arms, and an enormous rear wing, 40% larger than that of the previous generation. With a movable blade, this appendage, equipped with swan-neck masts, provided, for the first time on a production Porsche, the DRS drag reduction function. Depending on needs, the blade angled in or out, up to a maximum of thirty-four degrees. Not deviating from the RS formula, this model had its weight limited to 1,450 kg (3,197 lbs) with the use of carbon fiber reinforced plastic for the doors, front fenders, seat shells, and aerodynamic elements. Just by removing one gear, compared to the 992's seven-speed PDK transmission, 20 kg (44 lbs) was saved.

While the 2022 GT3 RS did not shy away from unleashing its 525 hp on open roads, it was more at home on a circuit's asphalt.

Under the rear hood, Porsche Motorsport experts integrated a new version of the 4-liter naturally aspirated six-cylinder engine with four valves per cylinder and a dry sump, designed to accept synthetic fuel. It had 525 hp at 8,500 rpm, a gain of 15 hp compared to the GT3 using new camshafts – with modified profiles – an individual throttle body intake system, a cooling radiator instead of the trunk, and rigid valve actuation derived from competition. The aerodynamic concept and engine served a particularly refined four-wheel steering chassis, mounted on Unibal ball joints and extensively adjustable from the steering wheel. The adaptive dampers adjusted in compression and rebound over nine positions and independently between the two axles. From the steering wheel, it was also possible to select one of the three driving modes, as well as adjust PSM and the PTV Plus differential lock.

For the first time, traction control (TC) was deactivatable. Phenomenal braking capabilities relied, at the front, on 408 mm (16.06 in) diameter discs and six-piston fixed calipers and on 380 mm (14.06 in) discs and four-piston calipers at the rear. Carbon-ceramic discs were part of the optional program. The braking system combined with soft-compound Michelin Pilot Sport tires (275/35 R20 at the front, 335/30 R21 at the rear) mounted on lightweight forged aluminum wheels with center lock. Thus equipped, the GT3 RS was 10.6 seconds faster than the GT3 on the Nürburgring Nordschleife, reached 100 km/h (62 mph) in 3.2 seconds, and had a top speed of 296 km/h (184 mph).

ENGINE: Flat-six
DISPLACEMENT: 4.0L / 243.9 in^3
POWER: 518 hp at 8,500 rpm
TORQUE: 343 lb-ft at 6,300 rpm
GEARBOX: PDK seven-speed
DIMENSIONS (L x W x H): 15' 0.0" x 6' 2.8" x 4' 4.0"
WEIGHT: 3,362 lbs
TOP SPEED: 184 mph

The enormous rear wing contributed to aerodynamic performance. Downforce reached 406 kg (895 lbs) at 200 km/h (124 mph); 860 kg (1,874 lbs) at 286 km/h (178 mph). Values doubled compared to the previous generation.

911 # GT3 R
2022

Following the launch of the 992 generation, the Motorsport department was tasked with developing a new competition version compliant with endurance race regulations, such as for the 24 Hours of Le Mans, with the GT3 category of various championships worldwide. Launched in 2022, the latest GT3 R made a strong impression with its widened body, refined aerodynamic package based on redesigned bumpers to optimize airflow, imposing side skirts, a large rear diffuser, and a rear hood topped by a voluminous wing.

To reduce its weight to 1,250 kg (2,756 lbs), Porsche applied a revision of its lightweight design principle. The majority of body elements were made of carbon, including the underbody. At its heart, the naturally aspirated water-cooled flat-six engine displacement increased to 4,194 cm3 to claim 565 hp. This engine was mated to a six-speed automated gearbox.

As was customary for a race car, the interior was entirely stripped. The body was bare; a digital instrument cluster replaced the dashboard. The various functions of the car were accessible from a carbon control panel located centrally and oriented towards the driver. The racing bucket seat was moved towards the center to improve mass distribution. The chassis settings were modified compared to a production 911, within the limits of the regulations. The dampers thus had five adjustment levels. The anti-roll bars were adjustable, and shims were installed on both axles.

BELOW: The 911 GT3 R RennSport offered a new driving experience to enthusiasts of track days.

ABOVE: With its body widened to the limits allowed by homologation, air intakes in the fenders, imposing side skirts, spectacular rear diffuser, and voluminous rear wing, the GT3 R paid attention to aerodynamics.

RIGHT: Drivers had to display gymnastic flexibility to get into a GT3 R equipped with body reinforcements and a roll cage that contributed to rigidity.

ENGINE: Flat-six

DISPLACEMENT: 4.2L / 255.9 in^3

POWER: 557 hp

TORQUE: N/A

GEARBOX: Six-speed sequential

DIMENSIONS (L x W x H): 15' 1.9" x 6' 8.3" x n/a

WEIGHT: 2,756 lbs

TOP SPEED: + 186 mph

911 DAKAR
2022

When they entered the Los Angeles Auto Show in November 2022, visitors were far from imagining what Porsche had in store for them. On its stand, the German manufacturer unveiled a version that revived the memory of the first 911 to win the Paris-Dakar in 1984. Begun with the previous 991 generation, the genesis of this 911 with an 80 mm (3.15 in) raised ground clearance was far from a long, quiet river. For ten years, Thomas Krickelberg, the head of 911 development, and his team feared a veto from a sales and marketing department unwilling to take risks.

Finally, Walter Röhrl had to share his enthusiasm with the brand's decision-makers after driving a prototype of this strange raised 911. But, as it was written that nothing would be easy for the engineers, even the choice of name gave them a hard time. The initial idea of naming it "Safari," in reference to the 911's exploits in the African event in the 1970s, failed due to the refusal of the Indian manufacturer Tata, owner of the name. After negotiations with the ASO – Amaury Sport Organisation – the company organizing the rally-raid named after the capital of Senegal, the name "Dakar" was chosen.

Based on the widened body of the 911 Carrera 4 GTS with four-wheel steering and active anti-roll bars, German engineers developed a model capable of tackling sand dunes, wading through a ford, or crossing a snowy forest path. The Dakar was equipped with the perfect off-road kit, including wheel arches highlighted by a black plastic strip, steel side sill protectors, front and rear underbody guards, and a partially faired underbody.

The 911 Dakar was enhanced with new driving modes: "Rally" for soft ground and "Offroad" for sandy and rocky terrain. Ground clearance can reach 191 mm (7.52in), provided you don't exceed 170 km/h (105 mph).

At Porsche, nostalgia was in full swing. With the Dakar, a lifted 911, the German manufacturer proves once again that the benchmark sports car boasts incredible versatility.

The 911 Dakar was the result of ten years of tenacity by German engineers. It was finally the former world rally champion, Walter Röhrl, who succeeded in convincing the brand's management.

It offered a better off-road approach and departure angle than the Cayenne, thanks to its short wheelbase. This performance was achieved when the long-travel suspension was in its high position. In this case, the ground clearance, already raised by 50 mm (1.97 in) compared to a Carrera with the sport chassis, gained another 30 mm (1.18 in) to reach 191 mm (7.52 in). A position it could maintain up to 170 km/h (105 mph). Beyond that speed, up to 240 km/h (149 mph), the speed allowed by the specific Pirelli Scorpion studded and reinforced sidewall tires developed for the car, the ground clearance returned to its initial position. Porsche neglected nothing: for those planning to go into the sand, sand recovery boards and a folding shovel could be paired with the optional roof rack.

Despite these modifications, the weight gain was limited to 10 kg (22 lbs) by using carbon fiber reinforced plastic for the front hood and fixed rear wing, but also reduced sound insulation, thinner glass, and the removal of the rear seats. Equipped with the 480 hp 3-liter twin-turbo flat-six boxer engine from the GTS and the eight-speed PDK automatic transmission, the 911 Dakar was a limited series of 2,500 examples.

By choosing the "roof rack" option, the 911 Dakar became a true adventurer. This equipment included sand recovery boards, a folding shovel, jerry cans, and a battery of spotlights.

ENGINE: Flat-six biturbo
DISPLACEMENT: 3.0L / 181.9 in^3
POWER: 473 hp at 6,500 rpm
TORQUE: 420 lb-ft from 2,300 to 5,000 rpm
GEARBOX: PDK eight-speed
DIMENSIONS (L x W x H): 14' 10.3" x 6' 1.4" x 4' 4.7"
WEIGHT: 3,538 lbs
TOP SPEED: 149 mph

GN-286-GY

911 GT3 R RENNSPORT 2023

What would a 911 GT3 R freed from all regulatory constraints look like? Porsche provided the answer in the form of a GT3 R RennSport unveiled at the Porsche International Meeting in late September 2023 at the Laguna Seca circuit in California. Apart from the hood and roof taken from the GT3 R, all carbon body elements were new. Truly impressive, the 911 RennSport was distinguished by a completely redesigned airflow and aerodynamic management. The new fenders integrated oversized air intakes; the rear was topped by an enormous wing whose shape recalled that of the 935/77 from the Brumos team seen at Daytona in 1978. The exterior mirrors were replaced by a system consisting of three cameras, two of which were integrated into the front fenders. An enormous carbon apron encircled the rear bumper, contributing to aerodynamic downforce.

Freed from all regulatory constraints and restrictions imposed by the balance of performance (BoP), the 911 GT3 R RennSport pursued a single objective: absolute driving pleasure. The 4.2-liter naturally aspirated flat-six engine from the GT3 R was upgraded to 620 hp at 9,000 rpm and could rev to 9,400 rpm due to specialized pistons and camshafts. With four valves per cylinder and direct fuel injection, this engine set a new record with a power output of 148 hp per liter. It was developed to accept E25 fuels, whether bioethanol or eFuel. Engineers were not content with simply taking the GT3 R's six-speed sequential gearbox to transmit the 620 hp to the rear wheels. The gear ratios for fourth, fifth, and sixth gears corresponded to the Daytona circuit configuration of the GT3 R race cars.

The chassis was identical to that of the GT3 R with a double wishbone suspension at the front and a multi-link suspension at the rear coupled with adjustable dampers in five positions. With the new 117-liter fuel tank and the removal of air conditioning, the RennSport managed to reduce its weight to 1,240 kg (2,734 lbs), which was 10 kg (22 lbs) lighter than the GT3 R. It was produced in a limited series of only 77 examples.

OPPOSITE LEFT: A true competition car, the GT3 R RennSport incorporated the most advanced aerodynamic solutions that relied on deflectors, slats, and active flaps.

OPPOSITE RIGHT: The 911 GT3 R RennSport offered an impressive driving experience to enthusiasts of track days.

ENGINE: Flat-six

DISPLACEMENT: 4.2L / 255.9 in^3

POWER: 612 hp at 9,000 rpm

TORQUE: N/A

GEARBOX: Six-speed automatic

DIMENSIONS (L x W x H): 15' 1.9" x 6' 8.7" x n/a

WEIGHT: 2,734 lbs

TOP SPEED: 211 mph

The GT3 R RennSport was the v sion of a race car freed from homologation constraints. Its rear wing recalled the 1973 935 Brumos, and its 4.2-l ter flat-six delivered 620 hp.

911 S/T 2023

At Porsche, it had become a habit that every 911 anniversary justified the marketing of a special edition model. Thus, in mid-August 2023, the Stuttgart manufacturer stirred excitement by unveiling the S/T version. Produced in a limited series of only 1,963 units, this model revived a label that designated competition 911s in the early 1970s.

The 2023 911 S/T combined the technical elements of the 911 GT3 RS with the body of a GT3 Touring. It was also the lightest model in the 992 family. The naturally aspirated 4-liter flat-six engine, delivering 525 hp at 8,500 rpm and 465 Nm (343 lb-ft) of torque at 6,300 rpm, was paired for the first time with the six-speed manual gearbox with shorter ratios. By adding a lightweight clutch combined with a single-mass flywheel, reducing rotating masses by 10.5 kg (23.1 lbs), as well as making the front hood, roof, doors, and bucket seat shells, roll cage, rear anti-roll bar, and rear axle reinforcement from carbon fiber reinforced plastic, not to mention forged magnesium center-lock wheels, carbon-ceramic brake discs, a lithium-ion battery, and lightweight glass, the S/T saved 40 kg (88 lbs) compared to a GT3 Touring.

Leaving efficiency to the GT3 RS, the S/T version dispensed with rear-wheel steering but retained the mechanical limited-slip differential. The chassis settings were revised accordingly. The aerodynamics relied on bumpers similar to those of the GT3 Touring and a retractable rear spoiler with automatic deployment.

The interior ambiance mirrors that of the GT3 Touring, with full bucket seats upholstered in leather along the sides and fabric for the seat centers. Like 1960s-era 911s, the tachometer and the Sport Chrono package stopwatch feature green markings. For this exclusive model, only two standard body colors – white and black – were offered, along with five special metallic finishes: Rivage Blue, Shade Green, Ice Grey, Gentian Blue, and GT Silver.

The 4-liter naturally aspirated flat-six engine, producing 525 hp at 8,500 rpm, was paired for the first time with the six-speed manual gearbox with shorter ratios.

Once again, Porsche drew on its sporting history to maintain the 911's appeal. The ST designation was associated with competition models from 1969 to 1972.

ENGINE: Flat-six
DISPLACEMENT: 4.0L / 243.9 in³
POWER: 518 hp at 8,500 rpm
TORQUE: 343 lb-ft at 6,300 rpm
GEARBOX: Six-speed manual
DIMENSIONS (L x W x H): 15' 0.0" x 6' 0.9" x 4' 2.4"
WEIGHT: 3,362 lbs
TOP SPEED: 186 mph

With the 911 S/T, Porsche continued its exclusive sports car concept. Limited to 1,963 units, this model combined the engine of the 911 GT3 RS with the body of a GT3 Touring.

911 CARRERA GTS LE MANS CENTENAIRE EDITION
2023

In 2020, designer Grant Larson, designer of the first Boxster and now head of special projects within the styling department, was approached by Porsche France to imagine a 911 that would pay tribute in 2023 to the German firm's return to the top endurance category. Holder of the record for victories at Le Mans (nineteen) and world endurance titles (sixteen), Porsche had decided to return to the top endurance category with the 963 Hypercar, in the year of the 24 Hours of Le Mans centenary. Led in collaboration with Porsche's "Exclusive Manufaktur" department and the Automobile Club de l'Ouest, this project resulted in a series produced in only 72 units, referring to the number of years of the German brand's uninterrupted presence at Sarthe. Reserved for the French market, the 911 Carrera GTS Le Mans Centenaire Edition was distinguished from a production vehicle by twenty-five specific elements combined with about thirty options from the catalog.

The Aluminum Gray paint referenced the first 356 number 46 that raced at Le Mans in 1951, driven by Auguste Veuillet and Edmond Mouche; the side quarter panels and wheels referred to the 911 GT1 with which Allan McNish, Stéphane Ortelli, and Laurent Aïello won Le Mans in 1998. The rear grille was adorned with a hand-milled badge as well as a 3D badge featuring the circuit layout in blue, white, and red.

Upon opening the doors, passengers were greeted by door sills bearing the model's monogram and Graphite Blue leather-velvet mixed upholstery. The reference to the Sarthe circuit was visible on the central armrest with the embossed track outline on the leather and the 24 Hours of Le Mans logo on the headrests.

Based on the 480 hp 911 Carrera GTS, the Le Mans Centenaire version was available, optionally, with a seven-speed manual gearbox or an eight-speed PDK gearbox. Each car was delivered with the ignition key in a case, a cover to protect the vehicle, and a personalized book detailing the different stages of its manufacture at the Zuffenhausen factory and in the "Exclusive Manufaktur" workshops.

Available in the year of the centenary of the 24 Hours of Le Mans, the special series Le Mans Centenaire Edition was equipped with either of the two transmissions from the catalog: PDK with eight gears or manual with seven speeds.

ENGINE: Flat-six biturbo
DISPLACEMENT: 3.0L / 181.9 in^3
POWER: 473 hp at 6,500 rpm
TORQUE: 420 lb-ft at 2,300 rpm
GEARBOX: PDK eight-speed / seven-speed manual
DIMENSIONS (L x W x H): 14' 10.5" x 6' 0.9" x 4' 3.2"
WEIGHT: 3,406 lbs
TOP SPEED: 193 mph

ABOVE: The racing number 46 was not there by chance. It was the number of the first Porsche entered at Le Mans in 1951, a 356 driven by Auguste Veuillet and Edmond Mouche.

BELOW: Conceived by Porsche's "Exclusive Manufaktur" workshop, the 911 Carrera GTS Le Mans Centenaire Edition was an initiative of the brand's French subsidiary.

911 TURBO 50
2024

Half a century of the 911 Turbo was celebrated with an anniversary version based on the latest Turbo S originally launched in 2019. Enthusiasts of limited-edition models would be a little disappointed when discovering the characteristics of this collector's edition bearing the Heritage Design line label. Furthermore, the manufacturer, who had a sense of symbolism, limited the number of examples to 1,974, the birth year of the 911 Turbo. These models were all identical, with the exception of the paint. While Turbonite metallic gray – referring to the 911 RSR Turbo pre-series car exhibited at the 1973 Frankfurt Motor Show – represented the standard color, with the Heritage Design package, the Jubilee Turbo was delivered in Aventurine dark green. Each customer could also choose a color from the standard palette, or opt for a bespoke shade through the "Paint to Sample" program.

Externally, the anniversary Turbo was recognizable by its specific center-lock wheels and its "Turbo" inscriptions on the fenders, as on the first 3-liter Turbo. However, it was possible to order the vehicle without the exterior inscriptions. The 1974 911 Turbo largely inspired the interior with tartan pattern fabric upholstery in white and gray on the doors and the center of the enveloping seats. A special chronograph created by Porsche Design accompanied the delivery of each example. Based on the 650 hp Turbo S, the Turbo 50 was equipped with a sport exhaust, a 10 mm (.39 in) lowered chassis coupled with PASM adaptive suspension, and a braking system with PCCB carbon-ceramic discs.

ABOVE: The subject of a limited series made in only 1,974 units, in reference to the year the 911 Turbo was born, the anniversary model was disappointing for many enthusiasts.

RIGHT: Intended to celebrate fifty years of the Porsche 911 Turbo, the anniversary model launched in 2024 reconnected with the tartan upholstery of the early models from 1974.

LEFT: Like the first 3-liter 911 Turbos produced in 1974 and 1975, the Jubilee version regained the "Turbo" badging on the rear fenders.

ENGINE: Flat-six biturbo
DISPLACEMENT: 3.7L / 228.5 in³
POWER: 641 hp at 7,200 rpm
TORQUE: 590 lb-ft
GEARBOX: PDK eight-speed
DIMENSIONS (L x W x H): 14' 10.5" x 6' 2.8" x 4' 3.3"
WEIGHT: 3,616 lbs
TOP SPEED: 205 mph

911 SPIRIT 70
2025

Porsche continued to combine past and present. Launched in 2020, the Heritage Design program corresponded to a line of four limited-series 911s reinterpreting historical decorations. The first, based on the Targa 4S, was presented in 2020. Unveiled in 2022, the second, named Sport Classic, was developed from the 911 Turbo. With its ducktail, six-speed manual gearbox, and design inspired by the transition between the 1960s and 1970s, this model paid tribute to the famous 911 2.7 Carrera RS. The third was revealed on April 23, 2025. Still bearing the grille of the Porsche "Exclusive Manufaktur" custom department, it was developed from the 992 Carrera GTS Cabriolet with a hybrid engine that had been introduced to the market in 2024. Its baptismal name "Spirit 70" referred to the finishes in vogue in Stuttgart in the late 1970s and early 1980s. The three stripes that Porsche popularized during this period to improve the visibility of fast cars on the highway were revived.

On the 992 "Spirit 70," these three black horizontal stripes ran across the body and continued onto the soft top. This decoration combined with racing roundels, retro badges, black windshield pillars, and a distinctive deep green body color: "Olive Neo." The customization program, however, allowed choosing another color from the palette. The wheels, painted in a deep gray with bronze nuances, were inspired by the Fuchs models of 911s from the 1970s. The car inherited the GTS style front bumper with ten vertical and active air flaps, faired underbody, and redesigned headlights with four light points.

The interior treatment was typical of late 1970s design. For this special edition 911, stylists revived the "Pascha" pattern introduced in 1978 on the first 928. Mimicking a checkered flag, this graphic covered the center of the seats, whose bolsters were upholstered in black leather, the door mirrors, and even the inside of the glove compartment. Two colors were available: olive green or gray and black. The 911 Spirit 70 came standard with the Sport Design Package and the green-graduated instrumentation of the first 911s.

The third model in the Heritage Design family, the Spirit 70 version, reintroduced the "Pascha" patterned upholstery first introduced in 1978.

Popular during the 1970s to improve the visibility of fast cars on the highway, the three-stripe decoration on the front hood was the main feature of the limited edition Spirit 70.

With this technology called T-Hybrid, the 911 GTS greatly benefited from the experience gained in endurance racing and with the 911 GT3 R Hybrid. The system was based on a new 3.6-liter flat-six engine. Ultra-compact, this 485 hp engine used only a single 11 kW (15 hp) electric turbo. The technology was complemented by a 40 kW (54 hp) and 150 Nm (111 lb-ft) permanent magnet synchronous electric motor housed in the PDK gearbox, as well as a 1.9 kWh battery placed under the front trunk. With a combined power of 541 hp, the GTS approached the famous 911 Turbo (580 hp). It was credited with a top speed of 312 km/h (194 mph) and dazzling acceleration, taking just 3.1 seconds to reach 100 km/h (62 mph). Its production was limited to 1,500 units.

ENGINE: Flat-six turbo
DISPLACEMENT: 3.6L / 219.1 in^3
POWER: 478 hp at 7,500 rpm + one electric motor (combined 534 hp)
TORQUE: 449 lb-ft
GEARBOX: PDK eight-speed
DIMENSIONS (L x W x H): 14' 11.2" x 6' 0.9" x 4' 3.4"
WEIGHT: 3,858 lbs
TOP SPEED: 194 mph

70
PORSCHE

The Spirit 70 is the first 911 in the Heritage Design series to feature a convertible body and the latest hybrid engine from the Carrera GTS.

ACKNOWLEDGMENTS

This is the sixth book I've written about Porsche, but the very first dedicated entirely to the 911. The idea of focusing on these rare, limited-production 911s really took shape during lively conversations with David Kings, who leads Glénat's fine automotive books program and whose catalog is such a treasure trove for car enthusiasts. I also owe a great deal to Marie-Mathilde Perchet, who guided the project with both professionalism and a wonderful sense of care.

I could never have brought this book to life without Porsche's tireless preservation of press kits and archives over nearly four decades. Just as important was the help of Jean-Marie Reisser, who took on the tough work of digging through family papers, checking facts, proofreading, and catching mistakes large and small. I'm equally grateful to Karine Reisser, whose patience, encouragement, and good humor carried the project through to the finish.

Along the way, many friends and colleagues offered support that made all the difference: Detlev von Platen, formerly head of sales and marketing at Porsche; Marc Meurer, managing director of Porsche France; Fayçal Elasri, who heads Press and Communications at Porsche France, together with his deputy, Floriane Gouyer. I also want to thank fellow enthusiasts and friends of the marque – Serge Bellu, Lucien-François Bernard, Marc and Pierre Gosselin, Régis Mathieu, Marc Ouayoun, and Valentin Kaiser – for sharing their passion and insights. And I hold a special memory of Stephan Roux, who managed Porsche's press relations in France from 1988 to 2003, back when the brand was still represented by Sonauto before becoming a full subsidiary.

Finally, I owe heartfelt thanks to Frank Jung and Jörg Thilow, who so carefully safeguard Porsche's archives.

PHOTO CREDITS

Cover: RM Sotheby's

8-9: Corporate Archives Porsche AG; 11: Archives SR; 12-13: Archives SR; 14-15: Corporate Archives Porsche AG; 14: Archives SR; 15: Archives SR; 16-17: Archives SR; 18: RM Sotheby's; 19: RM Sotheby's (top), Archives SR (bottom); 20-21: Corporate Archives Porsche AG; 20: Archives SR; 22-23: Archives SR; 24-25: Corporate Archives Porsche AG; 27: Archives SR; 28: Archives SR; 29: RM Sotheby's; 30: Archives SR; 31: Corporate Archives Porsche AG (top), Archives SR (bottom); 32: Archives SR; 33: Gooding & Company (top), Corporate Archives Porsche AG (bottom); 34-35: Corporate Archives Porsche AG; 34: Corporate Archives Porsche AG; 36-37: Corporate Archives Porsche AG, Corporate Archives Porsche AG, Corporate Archives Porsche AG; 38-39: Archives SR; 40: Archives SR; 41: Corporate Archives Porsche AG (top), Corporate Archives Porsche AG (bottom); 42: Corporate Archives Porsche AG; 43: Corporate Archives Porsche AG (top), Archives SR (bottom); 44-45: Archives SR; 46: Archives SR; 47: Corporate Archives Porsche AG, Corporate Archives Porsche AG; 48-49: Archives SR: 50-51: Corporate Archives Porsche AG; 53: Archives SR; 54-55: Archives SR; 55: Corporate Archives Porsche AG; 56 -57: Corporate Archives Porsche AG, Corporate Archives Porsche AG, Corporate Archives Porsche AG; 58-59: Corporate Archives Porsche AG, Corporate Archives Porsche AG, Corporate Archives Porsche AG; 60: Archives SR; 61: Corporate Archives Porsche AG (top), Archives SR (bottom); 62: Corporate Archives Porsche AG; 63: Archives SR; 64; Archives SR (top), RM Sotheby's (bottom); 65: RM Sotheby's; 66: RM Sotheby's (left), Archives SR (right); 67: RM Sotheby's; 68-69: Corporate Archives Porsche AG; 70: Corporate Archives Porsche AG (top), Archives SR (bottom); 71: Archives SR; 72: Archives SR; 73: RM Sotheby's (top), Archives SR (bottom); 74-75: Archives SR; 76-77: Corporate Archives Porsche AG; 76: Archives SR; 77: Corporate Archives Porsche AG; 78-79: RM Sotheby's; 78: Archives SR; 79: Archives SR; 80-81: Corporate Archives Porsche AG; 83: Archives SR; 84: RM Sotheby's; 85: Corporate Archives Porsche AG (top), Archives SR (bottom); 86-87: Archives SR; 88-89: Corporate Archives Porsche AG, Corporate Archives Porsche AG; 90: Archives SR; 91: Corporate Archives Porsche AG (top), Archives SR (bottom); 92: Archives SR; 93: Corporate Archives Porsche AG, Corporate Archives Porsche AG; 94: Corporate Archives Porsche AG; 95: RM Sotheby's (top), Corporate Archives Porsche AG (bottom); 98-99: RM Sotheby's; 100-101: Corporate Archives Porsche AG; 103: Corporate Archives Porsche AG; 104: Archives SR; 105: RM Sotheby's (top), Corporate Archives Porsche AG (bottom); 106-107: Dominique Fontenat; 108-109: Corporate Archives Porsche AG, Corporate Archives Porsche AG, Corporate Archives Porsche AG; 110-111: Corporate Archives Porsche AG; 110: Archives SR; 111: Archives SR; 112-113: Corporate Archives Porsche AG; 114-115: TIBO; 116: RM Sotheby's; 117: RM Sotheby's; 118-119: Corporate Archives Porsche AG; 121: TIBO; 122-123: TIBO; 124: Archives SR; 125: Archives SR (top), Corporate Archives Porsche AG (bottom); 126: RM Sotheby's; 127: RM Sotheby's (top), Corporate Archives Porsche AG (bottom); 128: RM Sotheby's; 129: RM Sotheby's (top), Archives SR (bottom); 130-133: RM Sotheby's; 134: RM Sotheby's; 135: Archives SR; 136-137: TIBO; 138-139: Corporate Archives Porsche AG; 140-141: Corporate Archives Porsche AG, Corporate Archives Porsche AG, Corporate Archives Porsche AG; 142-143: RM Sotheby's; 144: Archives SR; 145: Corporate Archives Porsche AG; 146-147: TIBO; 149: TIBO; 150-151: Archives SR; 152-153: TIBO; 154-155: TIBO; 156-157: Corporate Archives Porsche AG; 157: Corporate Archives Porsche AG, Corporate Archives Porsche AG; 158-159: TIBO; 160: Corporate Archives Porsche AG: 161: Corporate Archives Porsche AG (top), RM Sotheby's (bottom); 162-163: Corporate Archives Porsche AG; 162: Corporate Archives Porsche AG; 163: Corporate Archives Porsche AG; 164-165: TIBO; 166-167: TIBO; 168-169: TIBO; 170-171: TIBO; 172: Corporate Archives Porsche AG; 173: Corporate Archives Porsche AG (top), Corporate Archives Porsche AG (bottom); 174: Archives SR; 175: RM Sotheby's; 176-177: TIBO; 178-179: RM Sotheby's – Patrick Ernzen; 180: Corporate Archives Porsche AG; 181: Corporate Archives Porsche AG, Corporate Archives Porsche AG; 182-183: Archives SR; 184-185: Corporate Archives Porsche AG, Corporate Archives Porsche AG, Corporate Archives Porsche AG; 186-187: TIBO; 189: TIBO; 190: Archives SR; 191: TIBO; 192-193: Arthur Mathieu; 194-195: Corporate Archives Porsche AG; 195: Corporate Archives Porsche AG; 196-197: Corporate Archives Porsche AG; 196: Corporate Archives Porsche AG; 198: RM Sotheby's; 199: Archives SR; 200-201: TIBO; 202-203: TIBO; 203: Archives SR; 204-205: TIBO; 206-207: Corporate Archives Porsche AG, Corporate Archives Porsche AG, Corporate Archives Porsche AG; 208-209: TIBO; 210-211: TIBO; 212-213: Corporate Archives Porsche AG, Corporate Archives Porsche AG, Corporate Archives Porsche AG; 214: Archives SR (left), RM Sotheby's (right); 215: RM Sotheby's; 216-217: Thomas Cortesi; 218-219: Corporate Archives Porsche AG, Corporate Archives Porsche AG, Corporate Archives Porsche AG; 220-223: Corporate Archives Porsche AG, Corporate Archives Porsche AG, Corporate Archives Porsche AG.